The Rebirth of Polypropylene: Supported Catalysts

Edward P. Moore, Jr.

The Rebirth of Polypropylene: Supported Catalysts

How the People of the Montedison Laboratories Revolutionized the PP Industry

Hanser Publishers, Munich

Hanser/Gardner Publications, Inc., Cincinnati

The Author:
Edward P. Moore, Jr., 718 Cheltenham Road, Wilmington, DE 19808, USA

Distributed in the USA and in Canada by
Hanser/Gardner Publications, Inc.
6915 Valley Avenue, Cincinnati, Ohio 45244-3029, U.S.A.
Fax: (513) 527-8950
Phone: (513) 527-8977 or 1-800-950-8977
Internet: http://www.hansergardner.com

Distributed in all other countries by
Carl Hanser Verlag
Postfach 86 04 20, 81631 München, Germany
Fax: +49 (89) 98 12 64
Internet: http://www.hanser.de

The use of general descriptive names, trademarks, etc., in this publication, even if the former are not especially identified, is not to be taken as a sign that such names, as understood by the Trade Marks and Merchandise Marks Act, may accordingly be used freely by anyone.

While the advice and information in this book are believed to be true and accurate at the date of going to press, neither the authors nor the editors nor the publisher can accept any legal responsibility for any errors or omissions that may be made. The publisher makes no warranty, express or implied, with respect to the material contained herein.

Library of Congress Cataloging-in-Publication Data
Moore, Edward P.
The rebirth of polypropylene: supported catalysts: how the
people of the Montedison Laboratories revolutionized the PP industry
/Edward P. Moore, Jr.
  p. cm.
Includes index.
ISBN 1-56990-254-2
1. Polypropylene. 2. Catalysts. I. Montedison S.p.A.
II. Title.
TP1180.P68M66 1998
668.4′234—dc21   98-33641

Die Deutsche Bibliothek - CIP-Einheitsaufnahme
Moore, Edward P.:
The rebirth of polypropylene : supported catalysts : how the people of
the Montedison Laboratories revolutionized the PP industry /
Moore. - Munich : Hanser : Cincinnati : Hanser/Gardner, 1998
ISBN 3-446-19587-4

© Carl Hanser Verlag, Munich Vienna New York, 1998
Typeset in England by Techset Composition Ltd., Salisbury
Printed and bound in Germany by Druckhaus "Thomas Müntzer", Bad Langensalza

Foreword

On its face, this book is a history, a comprehensive and fascinating history of the discovery and technological development over four decades of polypropylene (and later myriad derivative polymers). It guides us through multiple generations of scientific and technological advances and the epochal discovery of the active $MgCl_2$ catalyst support system that has itself been exalted into more advanced generations. Each generation is carefully accounted for—its origin, the thinking and the research that attended its evolution, the managerial issues that impeded or enhanced technical progress, and its ultimate expression in new theoretical understanding of Ziegler–Natta catalysis, the design of ever more sophisticated catalysts, the expanded performance range of the products derived from them, the brilliant process engineering that translated laboratory discovery into commercial reality, and the ever improving economics of products and processes.

Technological developments as commanding as these never, of course, take place in a single institution, one laboratory of one company. The book adds to the accomplishments of the Montedison researchers those of many competing companies which, for the success it sought and then achieved, Montedison R&D had to surpass. That it did and still today usually does is well recounted in this book, the tension of competition adding to the zest of winning.

Prof. Giulio Natta, who discovered and characterized isotactic polypropylene in 1954, and who shared with Karl Ziegler the Nobel Prize in chemistry for his discovery, declared in 1960 that there was little to be gained from further research in polypropylene; that the future would see only refinements in the already elegant theme represented by his earlier pioneering work. Genius though he was, this history is a powerful refutation of that sense of scientific exhaustion, and a powerful lesson in the value of continued research, of the willingness to challenge and renew understanding, and to apply new insights for generation after generation of scientific and technological advances.

Today, the annual consumption of about 4 kg of polypropylene and closely related products for every human on earth is testimony to the fruits of that on-going technical renewal. This rich history reveals why Montell product and process technologies lead the competitive universe of polypropylene. Accordingly, it is not only a history; it is a compelling instruction for all technologically based industry.

Thus this book is more than an account of technical challenge, struggle, achievement and regeneration. It is an account of much more, some of which industrial leaders will surely recognize in their own careers and companies:

- The heroic struggles (and successes) of a small band of dedicated researchers determined to succeed in the face of a research-blind management;
- The liberating effect—indeed the accelerating effect—of equally heroic enlightened and supportive management;
- The force of basic science on the advance of technology. "When you have a practical problem . . . you (first) understand the basic phenomenon. You interpret, look for laws, and then go to the solution. Not trial and error, but interpretation! That is absolutely the key . . . a deep knowledge of theoretical phenomena. Then balance with the practical and commercial." (Prof. Paolo Galli, in the 1970s–80s, research leader and spiritual driving force at the Ferrara Research Center. From *Break-throughs!*, Rawson Associates, New York 1986);
- Openness to alliances, to external sources of complimentary technology, closure to the venom of the "not-invented-here serpent";
- Exemplary integration of the science and technology of catalysis with the science and technology of process engineering and, ultimately, with the calls of the market place;
- Alertness to the potential for entirely new families of products and processes whose forebears reside in this tale of polypropylene catalysis but whose value could not even have been dreamt of as recently as a decade ago;
- The indispensale powers of commitment, drive, cooperation, and shared values within any R&D team.

Thus this volume is more than a rich history of frustration and success, exaction and excitement in the demanding and rewarding field of supported Ziegler–Nata catalysts for polypropylene. It is an exemplar of the successful research for riches that all technologically founded companies seek.

Philip A. Roussel
Director Emeritus, Arthur D. Little, Inc.

Preface

First, thanks go to Montell for sponsoring this project.

Second, some thoughts about this book. The task of describing the development of supported catalysts and consequent process improvements coming from the Montedison laboratories seemed simple enough when initially described. As I became familiar with the details, retrieved mostly from the reports in Ferrara, Italy, the complexity of the events and the significance of the technical achievements became more apparent. Yet, in the end, the story was more about the research people and their vision, spirit, and determination, than about inanimate, though revolutionary, techological achievements. While difficult to capture, I hope the reader can experience to some degree the excitement of discovery intended to be conveyed through this publication. I have tried to present these developments, obviously based heavily on Montell's records and recollections, in a readable but technically and historically accurate account of those exciting times.

Third, some observations about Italians we have met during the writing process. While I was seeking the ingredients for this story in Ferrara, my wife Georgie and I made another discovery that provides a dimension that at first appears separate from this assignment, but may, on reflection, turn out to be related. We came to know and appreciate the enchanting spirit of Italy that flows from its people. Any interest in their activities was quickly met with enthusiastic appreciation, and a generous sharing. Their friendliness and warmth were always far greater than expected. We particularly value our new friendships with Mrs. Govoni and staff at the Annunciata; the two Leonardi at L'Oca Giuliva; the Tassinari family at Cornice; Silvana (and Marina) from Residence Ariosto; Marco and Massimo at Al Pescatore, in Tresigallo; Giovana at Les Griffes; Andrea at Lanzagallo, in Gaibana; Gabriella at Negozio FIS; Andrea and family at La Romantica, and Nicola and family at Smorfia. These people are all true ambassadors of the irresistible charm of Italy. If you visit Ferrara, we hope you get the opportunity to share our enjoyment of these delightful people. Is it possible that their attitude of cooperation and sharing was also an essential ingredient of Ferrara's outstanding research group? For me, the answer is obvious.

Finally, the assistance of many people at Montell's Ferrara lab was invaluable, but I wish to express a special gratitude to Tonino Simonazzi and Giuliano Cecchin for their personal help and guidance on this project.

E.P. Moore, Jr.

Contents

Part I

The PP Industry before Supported Catalysts

1　Unsupported PP Catalysts

1.1　Introduction

This document is about the spectacular development of the magnesium chloride-supported olefin polymerization catalysts, beginning in 1968, and continuing to the present. While the technical achievements are in themselves a major success story, even more amazing is how a small group of Montedison researchers led the polypropylene community into truly economical processes with their advances in supported $MgCl_2$ catalysts.

In the early 1950s, a farsighted Montecatini (which became Montedison in 1966) supported the research of Professor Giulio Natta, in the Polytechnical Institute in Milan. Professor Natta's discovery of crystalline PP in 1954 occurred with numerous Montecatini (MC) scientists at his side. The subsequent flow of new scientific discoveries into the MC labs was unprecedented, and led to immediate commercial results. In fact, the whole world benefited from the Ziegler–Natta catalyst developments.

Although in the 1960s the paths of Natta and MC began to separate, the people of the MC labs always considered themselves part of the unique organization that discovered PP with Natta. When difficulties or challenges arose, they responded with the confidence of true champions. They knew their strengths and capabilities, and were willing to risk their very careers on their positive results. This story demonstrates that their confidence was well placed.

The catalyst development was, without question, revolutionary. The extent of the growth is illustrated in Fig. 1.1, where a measure I have called the "Progress Factor," increased by four magnitudes in two decades. In this factor, polymerization activity is the primary contributor to its value, but any departure from fully isotactic polymer diminishes its value. Its definition appears as Eq. 1.1.

$$\text{Progress Factor} = (\text{Activity, kg PP/g Titanium})/(100\text{-IsoIndex}, \%)^1 \qquad (1.1)$$

The progress in supported catalysts was not only sufficient for design of a new, more economical PP production process, the original intent, but also expanded the range of materials under the PP umbrella, and fostered the development of new products and processes not even considered possible in the beginning.

It is difficult, even in retrospect, to appreciate the scope and significance of the developments following the 1968 discovery of the active $MgCl_2$ support for PP catalysts. In order to obtain a sense of the complexity and impact of those developments, described in detail in this publication, they are summarized in Table 1.1. The

---

[1]I have used "IsoIndex" instead of "II" as the abbreviation for isotactic index.

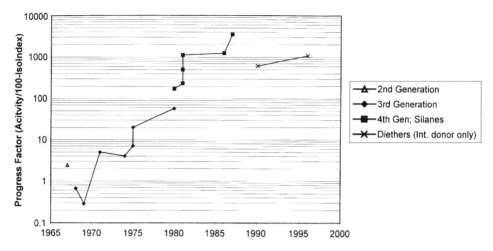

Figure 1.1   Progress in supported PP catalysts

developments up to the controlled catalyst particle size and shape in 1978 were anticipated prior to the successful realization of supported catalysts. Those after 1978 were mostly unexpected.

The developments after 1980, based on the Reactor Granule Technology, were only implemented following the recognition of the importance of porosity in 1984, although porosity played an important but not fully recognized role in the development of impact copolymers in 1981.

The atmosphere for this type of rapid and significant development was set early in Montecatini's history of PP. Not only did Montecatini have the vision to sponsor the research of Dr. Giulio Natta leading up to his first polymerization of PP in March 1954, but scaled up this new material, technology, and process to a commercial plant, which started up in October 1957. This was perhaps the most aggressive commercialization in history of an entirely new technology.

1.2   First Generation Catalysts

There are numerous excellent technical [1–11] and historical [12–14] summaries of the early catalyst developments following the preparation of high density polyethylene by Karl Ziegler in 1953, and isotactic polypropylene by Giulio Natta in 1954. Without dwelling on the details of those early catalysts, for which the reader should consult the above references, there were some aspects of the prevailing view of the chemistry and the catalyst behavior that influenced the development of the supported third generation catalysts.

Early work by Natta et al. and Choi et al. indicated that only the surface titanium atoms were sufficiently available to the aluminum alkyl to become active in the

Table 1.1 Supported PP Catalyst and Process Developments, 1968–1997

Year	Initial development	Technical result	Eventual process and economic effect
1968	Active $MgCl_2$ supported catalyst	Potential freedom to alter catalyst, polymer, and morphology independently	
1969	Internal donor	Higher IsoIndex, but lower activity	
1971	External donor	Very high IsoIndex, fair activity	
1975	Separate titanation	Good activity and high IsoIndex	Eliminate catalyst residues removal
1978	Controlled catalyst size and shape; Spherical support	High bulk density; Good slurry flowability; Large, uniform polymer particle size	Trouble-free processing
1980	Phthalate/Silane donors	High, sustained reaction rate, very high IsoIndex	Elimination of atactic removal and solvent recovery
1982	Spheripol process	Simpler plant layout	Lower investment and operating costs
1984	Porous catalyst	Tolerance to high rubber content	Reactor impact copolymers; Continuous low-blush copolymers
1985	Controlled, high porosity catalyst	Reactor Granule Technology	Catalloy process (and processes below)
1988	Increased rubbery phase content, wider composition range	Very soft copolymers	New supersoft products and applications from Catalloy process
1990	Non-olefinic polymerization on PP particle	Intimately dispersed graft copolymers	Hivalloy process
1994	Metallocene polymerization on PP particle	Multi-catalyst Reactor Granule Technology	Commercialization of metallocene PP

polymerization [15–18]. The Arlman and Cossee work showed that the titanium atoms at the corners, edges, and dislocations in the $TiCl_3$ crystals, having chloride vacancies, were the locations of the catalytically active sites [19]. As a consequence of these views, three approaches were considered to increase the amount of titanium accessible to the reaction, or obversely, to reduce the amount of inactive titanium:

1. Reduce the size of the $TiCl_3$ micro-particle, thus increasing the fraction at the surface
2. Use soluble transition metal catalysts
3. Support the $TiCl_3$ on a carrier.

1.3   Second Generation Catalysts

The success of the second generation (Solvay) catalysts were a direct result of approach (1) above [20–22]. The Solvay catalyst displayed a smaller primary crystallite size because Solvay found the conversion of the brown β-$TiCl_3$ into the active violet γ or δ form could take place at temperatures below 100 °C, instead of the 160–200 °C formerly used, because excess $TiCl_4$ catalyzed the transformation. At the lower temperature, smaller primary crystallites were formed. By combining this technique with an ether washing procedure to remove the $AlEtCl_2$ that normally formed during the catalyst preparation, and is a poison to the catalyst, the Solvay catalyst exhibited about a five fold increase in surface area, and a similar improvement in activity, while retaining isotacticity of about 95%. It is important to distinguish between the isotacticity measurement on the total polymerization product (t-IsoIndex) and that of the isolated polymer (p-IsoIndex), as much of the atactic material remains in the diluent, and thus increases the polymer IsoIndex. The total IsoIndex is a better measure of the catalyst behavior.

Details of the second generation catalyst composition and physical attributes appear in Table 1.2. The most complete yet concise description of the 2nd generation catalyst chemistry appears in Chapter 4 of Boor's book [1], as of about1973; the book published

Table 1.2   Typical Composition of Second Generation Catalysts (Late 1960s)

Analysis	Ti	27%
	Cl	65%
	Al	1.5%
	Amyl ether	4.5%
Polymerization	Yield (kg PP/g Ti)	10–20%
	t-IsoIndex	90–95%
Polymer analysis	Ti	45 ppm
	Cl	110 ppm
	Al	3 ppm

much later due to Boor's untimely death in 1974. Some improved forms of the Solvay catalysts are still being used commercially [23].

Some work was conducted at Montedison (ME) on approach (2) above, soluble catalysts, using benzyl derivatives of titanium and zirconium, which showed polymerization activity with ethylene and propylene, even without the use of aluminum alkyl cocatalysts [24]. Although that work demonstrated that stereospecific polypropylene could be prepared with homogeneous catalysts, the activity was extremely low, and the approach was not pursued.

However, approach (3), the attempts to place the $TiCl_3$ on a support, following delays, confusions, and detours, eventually returned achievements and rewards surpassing anyone's wildest dreams in the early 1960s.

2   PP Processes with Unsupported Catalysts

Prior to the development of supported catalysts, there was a great deal of activity aimed at simplifying the PP processes, although with limited success. The approaches usually involved conventional chemical engineering, principally aimed at devising more innovative and economic means of removing the catalyst residues and atactic polymer from the liquid phase. Many solutions were in place in the early 1970s [25, 26].

2.1   Slurry

The slurry process was complex, but versatile, as shown in Fig. 2.1, illustrating the continuous configuration, where the reactors were run full. Reactor temperature was controlled by jacket cooling. With the more active second generation catalysts, the unreacted monomer could be consumed in a react-down stage, as shown. Another option, necessary for less active first generation catalysts, was to flash off the monomer by lowering the pressure, recompressing, and recirculating. If a copolymer were being made, the faster-reacting ethylene was usually completely consumed in the reactor. Although four reactors are shown to illustrate the four reaction stages, many more reactors were often used in series.

In the work-up, following reaction quenching and solubilizing the catalyst with alcohol and water, the resulting aqueous stream contained the catalyst residues, and the atactic PP remained in the organic stream, the diluent. The clean up of these two streams was rather complex. The most common diluent was hexane, although several producers used kerosene for all of its earlier plants, and Shell built one plant using butane [27, 28].

The slurry process allowed production of a wide range of PP products. In particular, the high ethylene capability allowed high rubber impact copolymers to be produced, especially in the batch mode, and the work-up scheme tolerated high solubles. Hercules was able to market a copolymer with 15–20% C_2, or about 35% rubber. However, individual producers differed greatly in the rubber level that could be produced reliably. Consequently, the range of product types, as dictated by the operabilities of the plants, varied considerably among producers.

2.2   Bulk (Rexene, Phillips)

The principal advantage of the liquid propylene (bulk) process was higher output rate, but the plant design was more demanding because of the need for higher pressure to

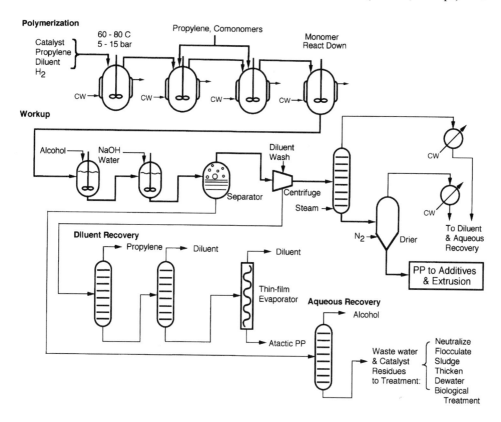

Polymerization

Catalyst
Propylene
Diluent
H₂

60 - 80 C
5 - 15 bar

Propylene, Comonomers

Monomer
React Down

CW → | CW → | CW → | CW →

Workup

Alcohol — | NaOH
Water

Separator

Diluent
Wash

Centrifuge

Steam →

CW

CW

To Diluent
& Aqueous
Recovery

N₂ → Drier

PP to Additives
& Extrusion

Diluent Recovery

Propylene → Diluent → Diluent

Thin-film
Evaporator

Atactic PP

Aqueous Recovery

Alcohol

Waste water
& Catalyst
Residues
to Treatment:

Neutralize
Flocculate
Sludge
Thicken
Dewater
Biological
Treatment

Figure 2.1   Slurry process (Hercules)

keep the monomer liquid. Beginning in the 1960s, Rexene and Phillips began polymerizing in liquid propylene, and developed clean-up processes that were simpler than that of the slurry process. The catalyst residues were rendered soluble in the organic phase by using an appropriate solvent. Consequently, the catalyst residues left the process in the same stream that contained the atactic PP, and ended up in the bottoms to the solvent still.

Rexene used an IPA/hexane azeotrope for the solvent, which simplified the distillation step. The Rexene process diagram appears in Fig. 2.2 [29–32].

In the Phillips process, the catalyst residues were treated with a "chelating agent," usually acetylacetone, plus an acid scavenger such as propylene oxide, and, along with the atactic polymer, were washed from the slurry by countercurrent contact with liquid propylene. The Phillips process is shown in Fig. 2.3 [33–36].

One limitation to the bulk process was the inability to have any significant concentration of ethylene in the reactor without forming a separate gas phase, which

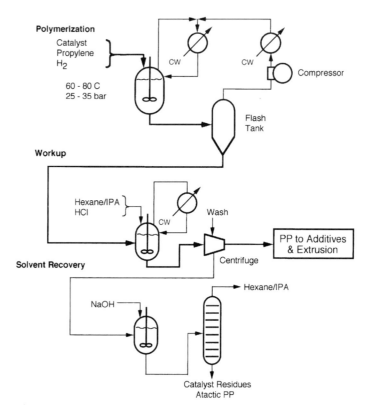

Figure 2.2   Bulk process (Rexene)

would have rendered the process inoperable. Thus, the achievable product range was limited to homopolymers and low ethylene random copolymers.

2.3   Vertical Stirred Reactor Gas Phase (BASF)

The vertical stirred reactor provided a very economic gas phase process, as shown in Fig. 2.4. Even with second generation catalysts, the BASF Novolen gas-phase process produced a "total" product; catalyst residues and atactic polymer remained in the product. Although the process was simple, the product was sensitive to color and oxidative instability from the catalyst residues, and the stiffness was lowered by the presence of atactic polymer. The acidic catalyst residues were neutralized with additives at the pelletizing extruder. The range of operable conditions was somewhat restricted by the process configuration, limiting the range of products capable of being manufactured.

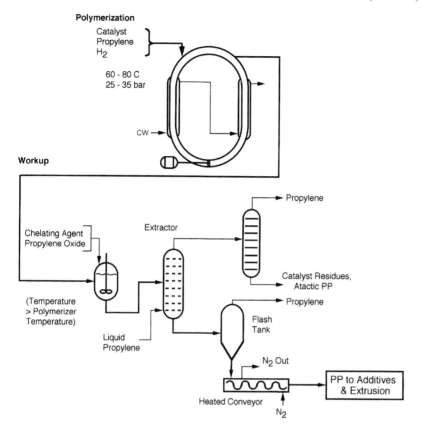

Figure 2.3   Bulk loop process (Phillips)

While the overall process was simple, the reactor was sophisticated. The specific design of the bottom-mounted agitator, which distributed the monomer, regulated the heat transfer, and mixed the polymer, was essential to the success of the process. Further, separate locations for the introduction of catalyst and cocatalyst streams were important.

The process development began in the early 1960s, and reached the pilot plant scale in 1967 at Ludwigshafen, and commercial operation in 1969 at Wesseling, Germany. A joint venture known as ROW (Rheinische Olefinwerke Wesseling) was formed with Shell to fabricate the Novolen plants for sale [37–42].

2.4   Solution (Eastman)

The solution process, illustrated in Fig. 2.5, was technically demanding, and only Eastman used it. This may have come about because of Kodak's extensive experience

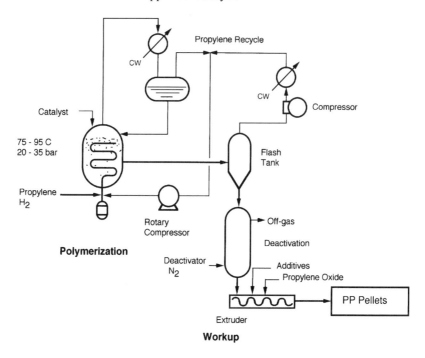

Figure 2.4   Vertical stirred gas phase process (BASF)

with, and thus confidence in, the polymer solution process, which they used in immense volumes to manufacture the cellulose acetate base for photographic film. The process allowed a moderate level of flexibility, although of a rather unique nature; the products tended to be softer and lower melting than products from the lower temperature processes. The product range was somewhat restricted because a special high temperature catalyst was necessary [28, 43–45]. This process is no longer used for producing crystalline PP.

2.5   Process Economics

While the development of the second generation (Solvay) catalysts improved the operating costs by increasing the activity by a factor of five, it did little to reduce the investment cost of a PP plant. Although it reduced or eliminated the need of removing the atactic polymer, this was of limited practical use, as the slurry plants still needed to distill the diluent, the bulk plants still removed the catalyst residues at the solvent still bottoms, and the gas phase plants still retained the atactic material. The lower atactic content was probably more important to BASF for the improvement in product

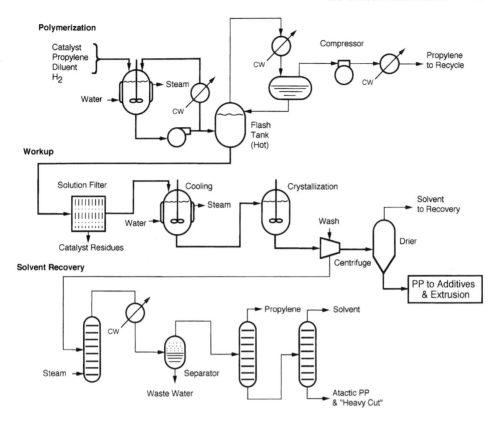

Figure 2.5   Solution process (Eastman)

properties it afforded; the product would be stiffer. All processes probably experienced a wider range of operable compositions.

However, at the dawn of the supported catalyst developments, little progress had been made in the reduction in PP plant complexity, and thus, construction or manufacturing costs.

Part II

The Revolution in Catalysts

3 Early Supported Catalysts

3.1 Concepts, Goals, Early Results

The concept of a supported catalyst was not new. In the early 1960s, the target levels were established that would constitute satisfactory performance for supported catalysts. With effective supported catalysts, specific compositional benefits were anticipated, as a direct result of the high activity (amount of polymer produced per unit of $TiCl_3$ catalyst), which translated directly into economic benefits [46]. For example:

- At a Ti level below about 5–10 ppm in the polymer, problems with yellow color development would disappear, and removal of catalyst residues would not be necessary.
- At a Cl level below 30–50 ppm, corrosion levels would be low enough that removal of catalyst residues would not be necessary to avoid corrosion. Also, note that triethylaluminum (TEA) must be used as the cocatalyst in such a system to avoid the corrosive effects of the easily hydrolyzed Al-Cl bond in diethylaluminum chloride (DEAC).
- The more critical requirement, the Ti content, would require an activity of at least 100 kg PP/g Ti. However, for PP, high activity was not the only objective. In order to maintain a satisfactory product, and take best advantage of the process economies that could result, other polymeric properties must not be compromised.
- It was necessary to maintain the isotacticity level of the PP to retain the property advantages derived from high crystallinity [47]. In particular, modulus, upper use temperature, hardness, barrier properties, and chemical resistance are all strongly influenced by crystallinity [48]. Beyond the property requirements, it was expected that at a sufficiently high isotactic level, above about 95% IsoIndex, the amount of atactic polymer would be low enough that the removal of the atactic polymer could be eliminated, provided a solvent-free process were used, and the small amount of atactic polymer could be left in the polymer. Thus, the high IsoIndex requirement satisfied a property need, and, in addition, could provide a process economy. The standard slurry/solvent process, however, would not share this economy, as there would always be some atactic polymer in the solvent that would require separation. Unfortunately, the separation cost would not be reduced with less atactic polymer, because the solvent is distilled away from the atactic residue. Therefore, the process savings would only be realized by abandoning the slurry process.
- In addition to isotacticity, molecular weight and MW distribution control needed to be adequate, at least as good as standard slurry polymers, for either PE or PP, to have marketable polymers.

With supported catalysts, a new capability would become feasible: controlling the size, shape, and density of the catalyst, and therefore, the polymer particle. Since the active catalyst would be carried just on the surface of the support, the support could be any shape desired, within physical and economic limits. Early work had shown that the shape of the polymer particle was determined, under the right conditions, by the shape of the catalyst particle; this behavior is known as "replication." Consequently, the most desirable shape for the polymerization and work-up process, as well as handling and downstream processing, could be considered. It was clear from even the most elementary considerations that a narrow particle size range and regular, preferably spherical, particles would be highly desirable, particularly for high reactor throughput and flowability, low entrainment, and high polymer bulk density.

The most desirable average particle size for the manufacturing process would be something larger than the usual slurry particles, and particles approaching the dimensions of extruded pellets would be most attractive for processing into extruded and molded products. For gas phase reactors in particular, larger particles free of fines would be most desirable to minimize entrainment during the gas/solid separations at the top of the reactor.

The density of the particle was, at that time, usually aimed at a high level to provide the maximum bulk density, throughput, and storage capacity. The potential benefits of a high porosity (thus, low density) particle had not yet been considered. However, the idea of density control was included in the considerations.

Therefore, the final criterion for good supported catalysts was:

- Control of the polymer particle size, shape, surface, and inner structure could be determined by the size, shape, and structure of the catalyst particle.

3.1.1   Early Experimental Results

Early attempts to support the $TiCl_3$ catalyst on "conventional" high-surface supports such as silica, alumina, magnesium hydroxides, or hydroxy-chlorides, while occasionally successful for polyethylene, were not effective for polypropylene due to very low levels of activity [49]. However, Solvay was able to reach some activity with propylene, but with moderate isotacticity, with $Mg(OH)Cl + TiCl_4 + AlEt_3$ [50, 51] and higher isotacticity was reached later [52]. The Solvay work quite consistently included an Mg-O bond in the supports [53].

3.2   The Discovery of Active $MgCl_2$

3.2.1   Montedison at Ferrara, Italy

The Ferrara location represents a curious confluence of forces. An ancient city, the walled section planned in 1492, Ferrara no longer shares the banks of the Po River as it did then, but retains a large portion of the original layout and character. The eastern dukedom prospered for centuries from the rich soil of the Po valley, the easy transportation afforded by the river, and the strong defense provided by the city walls.

The agricultural bounty influenced the location of the modern laboratories. The shortage of natural rubber during World War II precipitated a major effort to create a synthetic rubber, for which butadiene was needed. Lacking petroleum resources, Italy turned to a readily available material, sugar, from sugar beets grown in the Po valley. Fermented into ethanol, then dehydrolyzed into ethylene, from which butadiene could be prepared, the sugar beets became the source of synthetic rubber, at a major manufacturing facility at Ferrara. This installation was purchased by Montecatini after the war, and was converted to petrochemical raw materials. It became the foundation of the extensive Ferrara manufacturing facilities.

As elsewhere in the world, the development of synthetic rubber was given high priority at Montecatini in the 1950s. Consequently, one of the major projects that followed the Ziegler–Natta catalyst success was ethylene-propylene rubber (EPR). Ferrara was heavily committed to the manufacture of EPR and the downstream technologies suitable for its use in automobile tires. Polyethylene and polypropylene received comparatively little attention.

Unfortunately, by the late 1960s, the innovative Montecatini attitude that encouraged the Natta discovery and the early PP commercialization had been greatly diminished. The Montedison joint venture had formed in 1966, and the Edison people, with no polymer background and little interest in research, seemed to be in control. In the Ferrara lab, the management was assumed by people who lacked the technical background needed to defend and promote the research programs to top management. It became the task of the technical leaders within Ferrara to carry out the programs without top management approval.

3.2.2   The Discovery

In 1967 Montedison decided to purchase a polyethylene process from Phillips Petroleum. Prof. G. Crespi, then director of R&D for Montedison in Milan, was horrified that their company, after leading to the discovery of crystalline PP, would so easily relinquish the leadership in polyolefins. After reviewing Ferrara's technical position with key Ferrara people, Prof. Crespi challenged top management to allow the laboratory one year to develop their own PE process. The Ferrara scientists proposed that they aim for the next generation process, which they predicted would be even better than the Phillips process.

Having been given the year, work began in Montedison at a feverish pace to find a route to a PE process that surpassed the Phillips and Solvay approaches. Investigations proceeded simultaneously at several locations. By February 1968, curious results began to appear during the studies of the Solvay technology.

Solvay used MgO and Mg(OH)$_2$ as supports for the titanium compound believed to be the active catalyst, MgOTiCl$_3$, and claimed that total dryness was necessary to be effective. However, in the early work of Dr. Paolo Galli, investigating this approach, the results obtained were poorer than expected. The "trouble" was believed to be due to trace amounts of water in the MgO and solvents. Attempts to exhaustively dry the reagents resulted in even worse performance. Defying conventional wisdom, more water was intentionally added, and the catalytic activity immediately improved. It was clear at this point that a different chemistry than that proclaimed by Solvay was involved, and work proceeded on defining the new chemistry.

Solvay also prepared the active catalyst by boiling $Mg(OH)Cl$ with $TiCl_4$, and removing the unreacted $TiCl_4$ with frequent hexane washes. Thus, much of the $TiCl_4$ was wasted. Another Ferrara approach to developing an improved catalyst, taken by Dr. Adolfo Mayr, was to reduce the amount of $TiCl_4$ consumed by conducting the reaction with a more reactive titanium salt. The use of R-Ti-Cl_3 was tried, because of the extremely high activity of the R-Ti bond. $Mg(OH)Cl$ was contacted with R-Ti-Cl_3, obtained by reacting $TiCl_4$ with an aluminum alkyl at $-80°C$ and raising the temperature slowly to $20°C$, known as the "cold procedure." The resulting catalyst was very active with respect to Ti, but contained too much chlorine, due to the composition of the support.

In order to improve the activity relative to Cl, several trials were conducted, decreasing the Cl content in the support, but with the unexpected result that lower chlorine content gave poorer activity and vice versa. Upon testing pure $MgCl_2$, the catalyst was active. In this instance, the $MgCl_2$ was obtained by the removal of H_2O from $MgCl_2 \cdot 6H_2O$ at high temperature in an HCl stream. The resulting $MgCl_2$ contained many lumps that were eliminated by milling.

The positive result obtained with $MgCl_2$ suggested that there was a deposition of $TiCl_3$ from R-$TiCl_3$ rather than a reaction between -OH and $TiCl_4$. To test that hypothesis, comilling of $TiCl_3$ with $MgCl_2$ was tried, and, once again, the catalyst was very active. Thus, this investigation revealed a significant role for the active form of $MgCl_2$.

By using the comilling procedure, it became possible to prepare catalyst with an optimized content of Ti in order to maximize the mileage on Ti and total catalyst, providing a means of reducing the high chlorine content. The comilling procedure was also extended to $TiCl_4$ and other Ti salts and complexes, with positive results.

To understand the nature of the "active form," a commercial anhydrous $MgCl_2$, of fine particle size (thus requiring no milling) was used as a support. A catalyst was prepared by the "cold procedure," described above, but the activity was poor.

After an exhaustive analysis, it was determined that only the X-ray spectrum indicated the unique characteristic needed to activate the catalyst. When the catalyst was active, the $MgCl_2$ spectrum exhibited an X-ray pattern in which, in place of very sharp diffraction bands, broad bands or halos appeared. In the active form of the catalyst, $MgCl_2$ exists in a very disordered crystalline structure, which was originally obtained by milling. (More details regarding the active $MgCl_2$ structure appear in Section 4.5.1) Eventually, many other techniques were found to produce the same result.

Thus was formed the foundation for a long and exciting journey through the following decades: when employed as support for the Ti compounds, an active form of MgCl$_2$ provided unusually high polymerization activity, in this instance, for ethylene. Armed with this new information, the Ferrara laboratory was able, within the year allotted by the ME management, to achieve their first unbelievable feat: circumvention of the Solvay and Phillips patents, while at the same time, creating a whole new catalyst chemistry which was to become the foundation for the third generation of Ziegler–Natta catalysts [54–61].

3.2.3   Vision of the Future Potential

The process having thus begun, Dr. Galli promptly issued an internal Montedison report that considered the potential positive results that might be achieved [62]. Instead of limiting himself to the immediate technical targets, he projected the ultimate possibilities for the process and economics, looking well into the future, with confidence in the high potential for success he knew resided in the people of the R&D laboratories. The following translations from his 1970 paper, a "call to arms" delivered to the directionless, lethargic Montedison, illustrate the magnitude of the challenge presented to the Montedison research group:

> It is everywhere of particular technological interest to obtain polymers of elevated bulk density and with the absence of fine powder for the immediate response that such factors exercise on the economy of the whole process.
>
> Further interest arises from the possibility of obtaining the polymer in 'flake' form, to allow, in the limiting case, the elimination of the operations of strand extrusion and pelletizing, which from the high cost of investment and operation of the related equipment, influence and largely determine the cost of the product.
>
> Finally, the exact knowledge of the polymer growth mechanism in the catalyst granule and of the laws that govern it would permit clarification to a large degree of the already established phenomena in heterophasic polymerizations, with the possibility of intervening in the same to arrive in the end at a more exact definition of the process technology and improved product characteristics.

Thus, Dr. Galli strongly proclaimed that the Ferrara organization could accomplish far more than the rest of the company expected. As the history has demonstrated, more was accomplished than even he had expected.

Meanwhile, alternative approaches to generating the MgCl$_2$ from other precursors were developed; Solvay started with Mg alkoxides [63], Shell employed a Grignard source [64], and Stamicarbon used Mg alkyls, in a solution polymerization [65]. While it is the intent of this document to focus on the technical, not the legal and patent aspects of the developments, it is worth noting that most of the alternative routes to MgCl$_2$ fell under the definition of active MgCl$_2$ patented by Montedison. More details regarding catalyst developments appear in Section 5.8.

3.3   Application to Polyethylene

At the time of the discovery of active MgCl$_2$, Montedison was producing both PE and PP, and the initial application of this discovery was directed at PE.

3.3.1   Demonstration of Morphology Control

In his 1970 report, in addition to considering the long term economic goals of the catalyst developments, Dr. Galli explored the technical requirements for controlling the morphology of polymer from the supported catalyst, which anticipated the advances to occur almost two decades later. The replication phenomenon, the duplication of the catalyst shape in the polymer particle, had already been established in 1966 [66], and mechanisms for polymer particle growth had already been described in the scientific literature [67, 68].

The potential advantages of controlled morphology particles were broadly accepted:

- The higher bulk density of spherical particles of relatively narrow particle size distribution compared to wide particle size distributions from irregularly shaped milled catalyst particles would provide higher polymer content, better slurry flow, and less tendency to plug, thus giving higher productivity in commercial plants.
- In a gas phase process, elimination of excess fines would reduce problems with entrainment during separations.
- Larger particles could be handled and extruded like pellets, obviating the need for pelletizing extrusion.
- More consistent particle size would reduce variations in quality due to differences in diffusion, heat transfer, or fluid flow of the slurry.

Some of the difficulties of achieving the controlled morphology polymers were recognized as well:

- If too brittle, abrasion of the particles could result in undesirable fines.
- Any disintegration of the polymer particles due to uneven polymerization rates within the particle would prevent replication of the catalyst particle shape, and create more fines.

In the experimental part of his 1970 study, Dr. Galli prepared spherical support by spray cooling the molten complex MgCl$_2 \cdot$6H$_2$O, quite similar to that eventually used, along with several other Mg-based supports that proved to be less effective. He was able to confirm that the PE polymer particle growth patterns followed the mathematical models based on the physical processes occurring. That analysis helped predict when replication of the catalyst shape would occur in the polymer, and when disintegration of the growing particle from non-uniform growth would destroy the original shape. See Section 5.10.1 for details.

Among the results of the 1970 study was a list of requirements for replication that hold true as well today as they did in the late 1960s:

- Sufficiently high concentration of active centers
- Homogeneous distribution of active centers
- High specific surface and porosity
- Ready access of monomer to every point within the particle, and
- Sufficient support fragility to permit subdivision of the particle into primary crystallites during polymerization, but
- Sufficient integrity in the polymer to retain all parts of a single catalyst particle in a single polymer particle.

Considering the infancy of the active $MgCl_2$ technology, the analyses and foresight were amazingly accurate. Several other factors destined to grow in significance later were also discussed:

- Catalyst support surface area and porosity
- Catalyst titanium level
- Regularity of particles in size and shape
- Smoothness of particle surfaces, and
- Generation of fines from attrition during normal processing.

This study was prepared at a time when many skeptics in the industry considered morphological control to be either unworthy of attention, or unachievable. The report remains a fundamental reference with regard to replication and spherical supports.

In 1972, Dr. Galli confirmed in a Ferrara pilot plant the ability to prepare spherical PE from spherical catalyst support particles, using the same $MgCl_2 \cdot 6H_2O$ spray-cooled approach [69].

3.3.2   Elimination of Catalyst Residue Removal

With the active $MgCl_2$ support, the activity in ethylene polymerization became significantly higher than the traditional catalysts, especially based on Ti content, as shown in Table 3.1, which eliminated the need for catalyst residue removal [46].

The simplified process based on this active $MgCl_2$-supported catalyst was first used to prepare PE on a commercial scale at Montedison's Brindisi plant in 1971. The initial

Table 3.1   Catalyst Activity for PE (1971)

Catalyst	Activity (kg PE/(Mol Ti · MPa · h)
$TiCl_3 + AlEt_2Cl$	320
$TiCl_3 + AlEt_3$	710
$MgCl_2/TiCl_4 + AlEt_3$	17000

plant operation encountered serious difficulties with broken and fiber-shaped polymer particles. The research team was called in to help. It was deduced that the catalyst was too active, with too rapid growth of the particles early in the polymerization, causing disintegration of the growing particles. By reducing the catalyst milling time, which also reduced the activity, the polymerization rate was lowered slightly and the polymer growth took place once again in an orderly manner that maintained the particle integrity. Successful operation of the plant followed [70], and the Brindisi PE plant has continued to use supported catalyst to this day.

Other companies continued to pursue different catalyst forms. In his treatise on Ziegler–Natta catalysis, Boor cites (status as of 1973) numerous investigators using supports to polymerize ethylene, including SiO_2, MgO, Mg(OH)Cl, $MgCl_2 + MeOH$, $Mg(OEt)_2$, and $SiO_2 \cdot Al_2O_3$, using $TiCl_x$ or CrO_x compounds as catalysts [49]. In the same reference, an indication of the future direction of catalyst development appears, describing the use of $MgCl_2 \cdot ROH$ to chemically form the support, and a separate titanation step at Mitsui Petrochemical, which appeared in 1972 patents [71].

At Montedison, as well as in the rest of the world, many producers of Ziegler–Natta PE switched to the active $MgCl_2$-supported catalyst, eliminating the catalyst residue removal step with little difficulty. High activity supported Z–N catalysts have been widely used since the mid 1970s. Somewhat different than the PP business, the PE producers usually make their own catalyst on the plant site. Because each PE plant is usually designed to produce a particular PE type, and there is much proprietary tailoring of the catalyst to specific performance characteristics of the product, there is more local technology introduced into the manufacture of the catalyst [60]. However, the active $MgCl_2$ patent covered many of the producers of Ziegler–Natta PE. In addition, the spherical support concept continued to receive attention in PE, but it was in PP that it became a reality.

3.4   Application to Polypropylene

As a company, Montedison had more invested in PP than in PE, both in terms of capital and commitment. After all, this was the company that was working with Giulio Natta when he first prepared and characterized crystalline PP. Thus, although PE was technically easier to adapt to the supported catalyst, PP was potentially the more rewarding product for Montedison; ME operated four PP plants, and just one PE plant.

PP developments had not been ignored during the years following the active $MgCl_2$ discovery. However, the requirements for success with PP were a magnitude more difficult: the monomer was inherently less reactive than ethylene, and the polymerization needed to proceed at a high level of stereoregularity. Without the high stereoregularity in the polymer, the process economics would suffer from the need to remove the atactic component, and the most desirable product properties, dependent on high crystallinity, would be diminished. Thus, with the focus on PP, the need for high stereoregularity, while retaining the high activity, became critical.

Table 3.2 3rd Generation Catalysts without Donors (Milled/Slurry, 1969)

Catalyst	Activity, kg PP/g Ti	IsoIndex, %
2nd generation	12	95
3rd generation; no donor	35	45

Using the third generation supported catalysts that were effective in ethylene polymerization, the initial results with propylene were not encouraging. While a higher activity was reached, the stereospecificity was very low, as shown in Table 3.2.

3.4.1 World-Class Catalyst Expertise

Quite early in the development of an improved PP catalyst, better donor behavior was recognized as necessary. Incompletely understood, donors provided a means to raise the isotacticity, although at the expense of the activity. Thus, improvement in the catalyst behavior probably would require better understanding of the chemistry of the donors and catalysts. Fortunately, there was available a source of expertise, with deep roots in catalyst chemistry.

The roots began in the Polytechnical Institute of Milan where Natta was in charge of the Organic and Industrial Chemistry Laboratory. Several dozen people working directly for Montecatini were under the supervision of Natta and his associates when PP first appeared on the scene in 1954. One in particular, Prof. Umberto Giannini, co-authored early PP publications with Natta, and authored many significant Montecatini reports. He remains a valued consultant to Montell even today, and personifies the continuous line from Natta to the present lab at Ferrara.

Montecatini moved most of its Natta lab people to a private Milan lab in 1958. There, a core of people with direct connections back to the Natta group were in place when the question of applying the active $MgCl_2$ to PP was asked in 1968. The response was swift and focused.

3.4.2 Electron Donors

It was already known from the first and 2nd generation catalyst behavior that the level of stereoregularity in PP could be raised with the use of Lewis bases (electron donors), but at a severe cost in activity [72]. The electron donors preferentially deactivated the aspecific catalyst sites, but also some isospecific sites [73, 74]. From this starting point begins a fascinating part of the PP story: the development of effective electron donors and the processes for using them.

Starting in 1969, a wide range of electron donors was tested in a systematic evaluation of their effectiveness. The objective was to move the polymerization off of the normal relationship between activity and stereospecificity, where higher levels of

Table 3.3   Internal Donors in 3rd Generation Catalysts (Milled/Slurry, 1969)

Catalyst	Activity, kg PP/g Ti	IsoIndex, %
2nd generation	12	95
3rd generation; no donor	35	45
3rd generation; internal donor	10	65

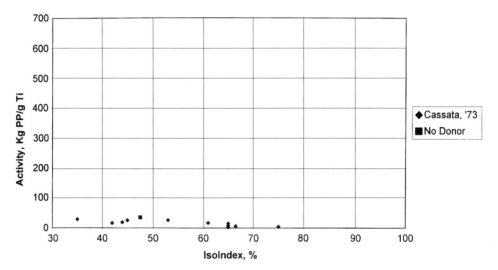

Figure 3.1   Performance of PP with internal donors.[1,2]

stereospecificity could be achieved only at the expense of very low activities. Some chemical families, such as the esters of aromatic carboxylic acids, were found to provide a modest increase in stereospecificity, with a small loss in activity. Thus, as with the unsupported $TiCl_3$ catalysts, the use of an electron donor in the catalyst system could improve the stereoregularity of the polymer, but not to an attractive level, and the activity was reduced, as shown in Table 3.3 and Fig. 3.1.

Analysis of the electron donor behavior revealed the following [75, 76]:

1. The electron donor was primarily coordinated with the $MgCl_2$
2. The donor was largely displaced from the $MgCl_2$ by TEA, a strong Lewis acid, present in high concentration

---

[1]The apparently inappropriate scale used in Fig. 3.1 will remain the same in future figures, to better illustrate the progress over time.

[2]Although the sources of the data in this series of figures are proprietary, the individual researchers and dates are included to indicate that the results identified by that symbol are from a single investigation.

3. A minor portion of the titanium goes into solution, and is available to form new active sites following displacement of the electron donor.

In addition to the above exchanges on the $MgCl_2$ coordination sites, some of the donors can undergo irreversible chemical reactions with the TEA, depleting the concentration of donor.

Two approaches were attempted to reduce the loss of electron donor:

1. Polymerize at low TEA concentration
2. Provide additional electron donor by including it in the TEA. The electron donor in the TEA became known as an "external" donor, and thus the electron donor in the $MgCl_2/TiCl_4$ mix became an "internal" donor.

Using the first approach above, low TEA concentrations were able to reach higher levels of isotacticity, but still below the level needed to eliminate the atactic polymer removal section. In addition, the results were inconsistent because the cocatalyst TEA acted as a scavenger for poisons in the polymerization mixture, and at the lower levels, minor fluctuations in the reaction mixture strongly affected the polymerization performance.

In a major development, use of the second approach above showed that employing an external donor gave both high IsoIndex, and retained an acceptable level of activity, as seen in Table 3.4 and Fig. 3.2. Initially the same Lewis bases, such as the aromatic acid esters, were used as both internal and external donors, but it was later learned that different donors could give better performance.

With the improvements from the inclusion of external donors, by the early 1970s the isotactic content and activity had reached the point where elimination of the catalyst and atactic removal sections from the process appeared feasible [77–82]. Work began on the catalyst preparation and polymer manufacturing, and downstream evaluation of the products obtained. By 1972, the ethyl benzoate (EB)/p-ethyl anisate (PEA) or /methyl p-toluate (MPT) combination was in place during a major pilot plant campaign to demonstrate high yield PP, using liquid monomer in a batch reactor, at Ferrara.

In that campaign, activities over 200 kg/g Ti were demonstrated using liquid monomer, which furnishes significantly higher activities than slurry, at IsoIndex

Table 3.4   External Donors in 3rd Generation Catalysts (Milled/Slurry, 1971)

Catalyst	Activity, kg PP/g Ti	IsoIndex, %
2nd generation	12	95
3rd generation; no donor	35	45
3rd generation; internal donor	10	65
3rd generation; internal and external donors	30	94

Note: Many of the examples used to illustrate progress in these tables were not obtained under the same conditions. Thus, while the overall improvements are evident, individual comparisons may not be valid.

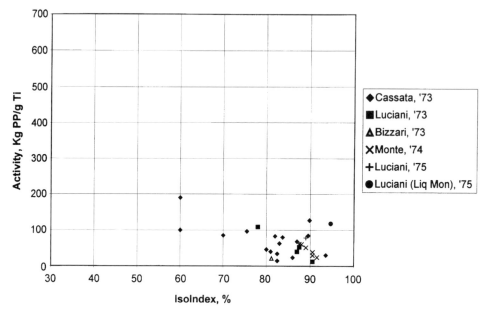

Figure 3.2   Performance of PP with external donors

values up to about 90%. However, the IsoIndex results were obtained on separated polymers (p-IsoIndex), and those values were higher than those associated with the whole polymerization (t-IsoIndex). Thus, more realistic values would be lower, but it is difficult to estimate how much from the reported results and methods. Yet, although the IsoIndex was still too low to allow elimination of the atactic polymer removal step, those results were very encouraging.

In these milled catalysts, the total titanium level could be varied at will by adjusting the catalyst composition fed to the mill. Initially, as Ti content increases, catalyst activity per gram of catalyst increases, although when expressed per gram of Ti, it would be almost constant, and IsoIndex level drops. At higher Ti levels, when the active sites are essentially fully occupied, the activity per gram of catalyst becomes constant with additional Ti, and per gram of Ti, becomes inversely proportional to Ti content. IsoIndex continues to drop. In this latter condition, each additional amount of Ti does nothing to activity, but simply adds ineffective Ti to the catalyst.

The most effective Ti level is thus a compromise between productivity per gram of catalyst, the cost of the added Ti, and the cost of its later removal from the polymer. In addition, the IsoIndex is affected by Ti level, in ways that depend on the catalyst preparation conditions. At very low Ti levels, less than about 0.2%, the catalyst is entirely aspecific [83], so a further compromise is sought to balance activity against specificity. In actual practice, about 1.7–2.5% Ti was eventually used commercially in milled catalysts.

It has been observed that the concentration of titanium plus the internal donor remains relatively constant over a rather wide range of conditions [74]. This is interpreted as indicating that the $TiCl_4$ and internal donor are strongly complexed to the available sites on the $MgCl_2$ support, and compete for those sites.

In the years immediately following the 1972 trials, little progress in activity or IsoIndex was made, as indicated in the dates of the investigations listed in Fig. 3.2. Some of the lack of progress was because Montedison did not have an adequate PP pilot plant for scaling up the process, and, perhaps more important, the commitment to PP by management was absent.

Management apathy was exemplified in the closing of the Milan laboratory in 1973 and moving the people to the Donegani Institute in Novara, the site of Montedison's central research facility. As with the earlier transfer of personnel to the Milan laboratory from the Polytechnical Institute, some of the people preferred not to move, and additional connections to the Natta heritage were lost. Still, the identification with the Natta discoveries remained a proud and inspiring element at the heart of the Montecatini research organization.

3.4.3   Energy Crisis

Contributing to the difficulties facing management was the worldwide energy crisis that struck in 1973. Besides disrupting the profitability of all petrochemical companies like Montedison by increasing the raw material costs, the cost of steam and electricity suddenly became a major portion of the manufacturing cost of polypropylene. The distillation section consumed large quantities of steam, especially when difficult azeotropic separations were necessary, as in the aqueous recovery section of the slurry process. Thus, the economic incentive to eliminate the workup section of the plants, distillation in particular, rose dramatically with the sharp increase in the cost of energy.

3.4.4   Montedison, Mitsui Petrochemical Agreement

The loss of management direction that occurred following the formation of Montedison deteriorated further. In fact, for a period in the early 1970s, the research group at Ferrara was reporting through the Operations management. When in 1970 Dr. Galli suggested that Ferrara pursue an adequate PP catalyst, top management declined [84].

Regardless of these impediments, the work continued to be initiated and directed by the Ferrara technical leaders. In the search for improvements in the catalyst system, the developments at Mitsui Petrochemical (MPC) became of interest. Kashiwa at MPC had patented a process that separated the titanation step from the milling of the catalyst support and internal donor [71, 85]. Evaluations of that approach indicated that a significant improvement in activity was achievable.

MPC, also seeking improvements, was aware that ME held the patent on the most basic step to high activity, the activated form of $MgCl_2$. There was sufficient overlap between those catalyst technologies that both companies were opposing each other's patents. Regardless of the patent battle, interest in exchanging technologies between ME and MPC grew. With the cooperation of Dr. Italo Trapasso in ME and Dr. Yasuji Torii at MPC, an agreement to exchange research, development, production, and marketing information on PP, to cross-license the existing technologies, and to jointly develop them further, was reached in 1975 [84].

4 Third Generation Catalysts—The Breakthrough

4.1 Separate Titanation and Milled Catalysts

The benefits resulting from the ME/MPC agreement were immediate. A milled supported catalyst component, designated "FT-1," was the first developed under the joint technologies, and was the principal commercial catalyst for several years. The manufacturing process was as follows:

1. Mill $MgCl_2$ with internal donor (usually ethyl benzoate)
2. Titanate with $TiCl_4$ at high temperature
3. Wash with boiling heptane
4. Dry
5. Polymerize with $AlEt_3$ containing the external donor (usually p-ethylanisate (PEA) or methyl p-toluate (MPT)).

Key features of the optimized process included carrying out the titanation in a separate step, and titanation at a high temperature. This provided a substantial improvement in performance as indicated in Table 4.1. The more active catalyst was patented jointly by ME and MPC [86, 87].

The high temperature separate titanation procedure was able to dissolve more of the $TiCl_4 \cdot EB$ complex that otherwise caused, through its formation, the removal of active $TiCl_4$ from the support, with a loss of activity.

The FT-1 catalyst performance approached those critical levels of activity and IsoIndex that could eliminate the need to remove the atactic polymer and catalyst residues; activity over 100 kg PP/g Ti and t-IsoIndex over 95% were the crucial requirements. Consequently, FT-1 was quickly advanced through the pilot plant trials to a commercial scale trial. The commercial scale trial took place at the Ferrara plant,

Table 4.1 Separate Titanation in 3rd Generation Catalysts (Milled/Slurry, 1975)

Catalyst	Activity, kg PP/g Ti	IsoIndex, %
2nd generation	12	95
3rd generation; no donor	35	45
3rd generation; internal donor	10	65
3rd generation; external donor	30	94
3rd generation; separate titanation (1975)	100	95

Figure 4.1 Photo of PP from FT-1 catalyst (30X)

and simultaneously at MPC, in 1976. Those evaluations indicated somewhat lower activity than in lab trials, and a marginal IsoIndex level. By varying the polymerization conditions, an activity level that clearly eliminated the need for catalyst removal could be reached, at the cost of lower IsoIndex; atactic removal continued to be required. In addition, the particle size distribution remained quite wide, as indicated by the photograph in Fig. 4.1. Yet, these results were very encouraging.

4.2   The World Awakens

The revelation of significantly improved catalyst performance with the issuance of the ME/MPC patent on separate titanation in 1976 [86, 87] woke up the research world to the realization that a supported catalyst with sufficient activity and stereospecificity to permit the elimination of the catalyst and atactic polymer removal sections of the process was rapidly approaching. Consequently, all parties with an interest in catalyst systems renewed their activities in supported research. The approaches to research varied considerably. Some were content to explore improvements over the art revealed in the ME/MPC patents. Others concentrated on circumventing that art, to avoid the expense of paying royalties. Still others simply tested the new systems as they were disclosed to determine which they would license. Regardless of the approach, the overall effect on catalyst research was dramatic: activity increased sharply, and continued for many years. Further specific details may be found in Section 5.8.

4.3   Milled Catalyst Improvements

Meanwhile, fairly detailed work on FT-1 was conducted to study the effects of manufacturing process, equipment design, and operating conditions on catalyst performance. By 1980, the performance of FT-1 had reached a new level, as indicated in Table 4.2 and Fig. 4.2.

Characteristics of typical FT-1 catalyst, as of 1979, appear in Table 4.3.

4.3.1   Commercial Run with FT-1

A commercial campaign was conducted at Brindisi in 1978, where the overall process and product results were as predicted from the 1976 plant trial. As expected, it was possible to eliminate the removal of catalyst residues, but the atactic polymer, although at a reduced level, still had to be removed.

Table 4.2   Separate Titanation in 3rd Generation Catalysts (Milled/Slurry, 1980)

Catalyst	Activity, kg PP/g Ti	IsoIndex, %
2nd generation	12	95
3rd generation; no donor	35	45
3rd generation; internal donor	10	65
3rd generation; external donor	30	94
3rd generation; separate titanation (1980)	260	95.5

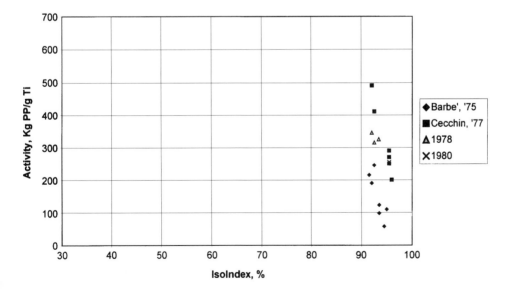

Figure 4.2   Performance of PP with separate titanation

Table 4.3   Typical FT-1 Catalyst Composition and Performance (1979)

Analysis	Ti	1.7%
	Mg	20%
	Cl	64%
	EB	7%
Polymerization[1]	Yld	250–300
	(kg PP/g Ti)	–
	t-IsoIndex	94–97%
	bulk density	0.45–0.50
Polymer analysis (calculated)	Ti	3 ppm
	Mg	40 ppm
	Cl	125 ppm
PSD of polymer	>1000 microns	17%
	>420	26%
	>177	30%
	>105	12%
	>53	13%
	<53	2%

[1]Conditions not reported; believed to be bulk, 70°C, 2 h.

In addition, the product exhibited an unacceptably high level of odor. Although difficulties with odor had been noticed earlier, the seriousness of this problem only became apparent to all levels of management with the arrival of tonnage quantities of malodorous PP from Brindisi. The call went out for an all-out assault on odor. Paradoxically, this difficulty initiated a concentrated effort, leading to another unexpected, but outstanding, success.

However, we shall postpone for a moment discussion of the events following the odor problem to review the status of the third generation catalyst and process concepts, and the initial work on the fourth generation catalysts and processes.

4.4   Process Status

4.4.1   Montedison Simplified Slurry

Taking advantage of the high yield obtainable with the milled catalyst FT-1, Montedison designed a process that eliminated the catalyst removal steps and the associated equipment. The aqueous stream, and all of the side streams it created, were entirely eliminated. This "simplified slurry process," illustrated in Fig. 4.3, was announced in 1976 [88, 89]. Recognizing that the atactic polymer content was not yet low enough to eliminate its removal, and considering the results in Fig. 4.2, the most economical plant would employ the catalyst with the highest activity, about 500 kg/g Ti. This approach gave the lowest catalyst cost, and the highest product quality with regard to catalyst

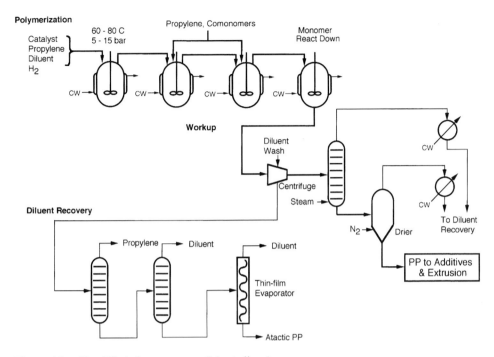

Figure 4.3   Simplified slurry process (Montedison)

residues. Mitsui Petrochemical offered a similar process in 1977 [90]. Commercial operation of the simplified process at the Montedison's subsidiary, Novamont, at the plant in La Porte, LA, demonstrated its economic advantage. The simplified process was reported to provide an advantage of 5–7 cents/pound in selling price over the next best process, liquid bulk, and more over the other liquid based processes [91]. The supported catalyst work was beginning to pay off.

4.4.2   Other Processes

As mentioned in Section 2, the bulk processes then in use (Rexene, Phillips) had already revised the chemistry of the catalyst residues to allow them to be removed in the same stream with the atactic polymer. Ironically, this meant that the elimination the catalyst residue removal changed their process flowsheets little, as the atactic removal stream still needed to be operated, and it used the same equipment. Therefore, the principal PP process to benefit from the simplified process was the old slurry process.

However, El Paso (Rexene) became interested in the more active ME catalyst, converted a PP plant to PE, operated it with the ME catalyst, and published the positive operating results jointly with ME [92].

The BASF gas phase process did not change, but the product quality could be improved by lowering the level of the catalyst residues; the stability and color problems could be reduced.

However, even the limited result with the simplified process demonstrated the power of the approach being taken by the Montedison/MPC coalition, and alerted the world that the PP catalyst technology was moving rapidly toward a breakthrough that had the potential to revolutionize the industry.

4.5   Supported Catalyst Concepts, Late 1970s

Thus, the immediate result of MPC agreement, in 1976, was a milled catalyst with adequate activity and IsoIndex for commercial application to the slurry process, with savings associated with the elimination of catalyst removal.

However, the elimination of atactic polymer removal had not yet been achieved, nor had any of the morphological improvements proposed in 1970 been implemented. Further, concern was growing about the inability to manufacture an effective impact copolymer using FT-1 in the slurry process. Nor could FT-1 be used in a gas phase process because of the excessive quantity of fines. In summary, the catalyst was not quite good enough for a major change in process, although its use in the simplified slurry process provided some improvement in the economics.

The greater challenges still lay in the future. Yet, much knowledge of the catalyst composition and behavior had been gained. In particular, the structure of the active form of $MgCl_2$, and the chemistry of the donors were much better understood, and are described briefly below.

4.5.1   Active $MgCl_2$ Structure

The key to the success of $MgCl_2$ as a support is the crystal structure, which is very similar to that of $TiCl_3$ in both interatomic distances and crystal forms. The detailed structure of the active form of $MgCl_2$ has been described well by Bassi *et al.* [93]. A simplified version follows:

The $MgCl_2$ crystal consists of layers of Mg ions sandwiched between two layers of Cl ions, as shown in Fig. 4.4. The arrangement of ions within this triplet layer of Cl-Mg-Cl remains the same among the several crystal forms. However, the spatial arrangement of the Cl ions from the relative placement of adjacent triplet layers defines the different crystal forms. The shaded Cl ions at the edge of the crystal in Fig. 4.4 may be used to illustrate. They occupy the "a" and "b" positions of a cubic close-packed array. If subsequent triplet layers continue the cubic array, as is the case for the α-form of $MgCl_2$, the Cl ions in the next layers occupy the "c" position, then the "a" position again, as shown in Fig. 4.5. Thus, α-$MgCl_2$ is described as having the ABCABCABC arrangement of Cl ions in a cubic close packed array, which gives a sharp X-ray reflection (104) at $d = 2.56 \text{ Å}$.

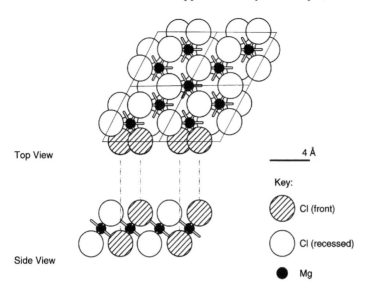

Top View

4 Å

Key:

Cl (front)

Cl (recessed)

Side View

Mg

Figure 4.4   MgCl$_2$ crystal triplet layers

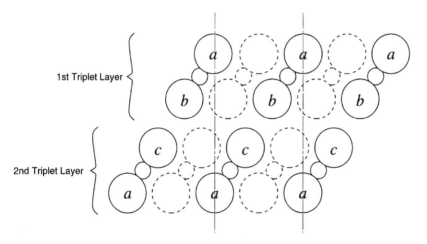

1st Triplet Layer

2nd Triplet Layer

Figure 4.5   Arrangement of Cl ions in α-MgCl$_2$

In β-form MgCl$_2$, the Cl ions are in a hexagonal close packed arrangement, and follow an ABABAB sequence, as shown in Fig. 4.6. This form exhibits a sharp X-ray reflection at $d = 2.78$ Å. However, in the active, or δ-form of MgCl$_2$, the stacking of the Cl-Mg-Cl triplet layers is disordered; both translation and rotation of layers occur. In δ$_1$-MgCl$_2$, where the succeeding layers are only translated, adjacent layers randomly assume the cubic or hexagonal forms. In δ$_2$-MgCl$_2$, where rotation of the layers has also occurred, the crystals are further disrupted. As a consequence, the X-ray reflection on

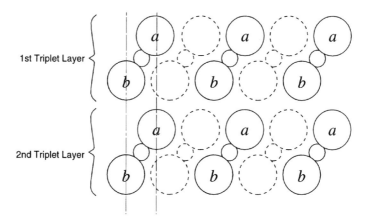

Figure 4.6   Arrangement of Cl ions in β-MgCl$_2$

the (104) plane becomes very broad, and is referred to as a "halo," at about $d = 2.65$ Å. While this broadening and that of the (003) reflection indicate loss of order through the crystal thickness, the broadening of the (110) and (101) reflections indicate that decreases have also occurred in the width of the crystallites.

Bassi et al. have described the X-ray patterns of the active MgCl$_2$ in terms of the types of crystals. Effective models for determining the factors responsible for the X-ray patterns, and thus, the polymerization performances, have been constructed [94–96]. Examples of the X-ray patterns of α-MgCl$_2$ and the activated δ-MgCl$_2$ appear in Figs. 4.7 and 4.8, respectively [97].

Only the breakage of the crystal on the (110) or (100) faces can create coordinatively unsaturated Mg atoms capable of forming catalytically active sites [98]. Thus, although high activity has been associated with the quantity of rotational faults in the δ_2-MgCl$_2$ crystal [99], it appears that this measure of disruption in the inter-layer Cl stacking, the most disordered of the α, β, δ_1, or δ_2 crystal forms, reflect the severity of the overall crystal disruption, which would be expected to relate to the reduction in crystallite size and the number of new (110) or (100) faces. It is this latter factor that would be expected to determine the potential for high catalytic activity.

4.5.2   Catalyst and Donor Chemistry

The chemistry of the third generation catalysts was elegantly summarized in a landmark review by Barbé et al., in 1987 [98]. Several other publications, while less complete, also provided useful descriptions of the catalyst behavior, and may be preferred by readers less expert in catalyst chemistry [100–102]. Of course, all the authors suffered from the usual difficulty in a developing technology: widely differing views about mechanisms and fundamental causes of observed phenomena can persist long after the behavior is described and even commercialized. The following in an attempt to simplify the description of the technologies reported in the above references.

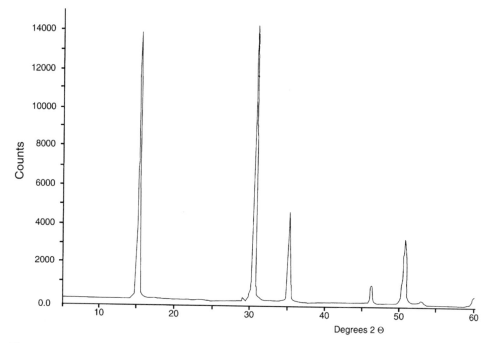

Figure 4.7   X-ray pattern of α-MgCl$_2$

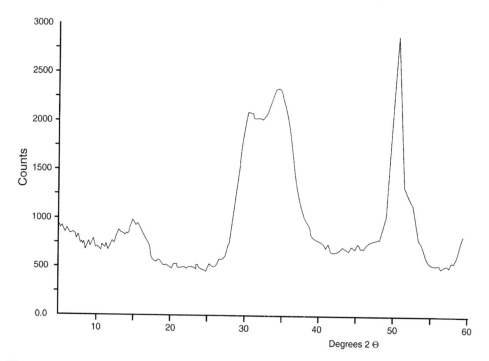

Figure 4.8   X-ray pattern of active δ-MgCl$_2$

4.5.2.1 Catalyst Preparation

MgCl$_2$ – TiCl$_4$ Interactions. When milled together, TiCl$_4$ formed a strongly bonded complex with MgCl$_2$, as evidenced by its retention in spite of repeated washing with solvent. During milling, the presence of TiCl$_4$ accelerated the activation of MgCl$_2$. Modification of the newly formed surface was believed involved; lubrication of adjacent crystals and prevention of re-agglomeration both seem to play roles. The titanium that was consequently complexed on the MgCl$_2$ was considerably more active than the Ti at the surface of second generation catalysts. Also, a higher concentration of complexed titanium, compared to Ti from addition of TiCl$_4$ to the milled MgCl$_2$, reflected the greater concentration of potentially active polymerization sites in the MgCl$_2$ caused by the presence of TiCl$_4$ during milling.

MgCl$_2$ – Internal Donor Interactions. Most of the reviewed work was conducted with EB as the internal donor. Complexation between EB and MgCl$_2$ in a still environment occurred in two stages:

1. A rapid adsorption of EB on the MgCl$_2$
2. The gradual formation of an EB $\cdot$ MgCl$_2$ complex.

The existence of the complex was evidenced by the shifting of the C=O IR band in the EB from 1720 cm^{-1} to about 1685 cm^{-1}. As the complex formed, the MgCl$_2$ crystal size was reduced.

During milling, the EB accelerated the MgCl$_2$ activation and reduction in crystal size, similar to the action of TiCl$_4$. This behavior was very sensitive to the EB/MgCl$_2$ mole ratio.

Interactions Between TiCl$_4$, and MgCl$_2$ and EB Milled Together. In this situation, the milled EB/MgCl$_2$ support was treated with TiCl$_4$ at a high temperature, and excess TiCl$_4$ removed by washing (Separate Titanation, Section 4.1). The TiCl$_4$ replaced some of the EB complexed to the MgCl$_2$ surfaces. Several advantages were observed (compared to omitting EB or adding it to the MgCl$_2$ after milling):

- Increased catalyst surface area (= smaller MgCl$_2$ crystals)
- Increased polymerization activity, about 3 ×
- Increased EB content, about 10 ×
- Higher Ti content (= more active sites), about 3 ×
- Higher IsoIndex, 92% vs 80%.

4.5.2.2 Polymerization

The kinetic behavior of polymerization with an early FT-1 catalyst was examined at Montedison in 1977. The reaction was characterized by a high initial rate, followed by a rapid decay, found to be typical of 3rd generation catalysts. The research established that the decay was not caused by diffusional considerations, but was due to a chemical change in the catalyst. As such, the decay was not affected by the amount of polymerization; it continued even while the monomer supply was interrupted.

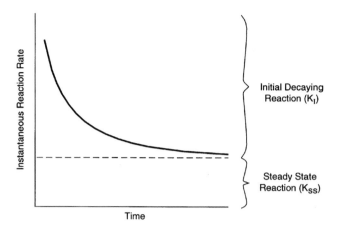

Figure 4.9   Polymerization kinetics: decaying and steady state

Different interpretations regarding the relationship between activity and IsoIndex were resolved in the above study when the FT-1 polymerization placed the activity/IsoIndex in a narrow band; within a given catalyst system, one could be improved only at the expense of the other (See Cecchin data, Fig. 4.2). Also, both activity and IsoIndex reached a maximum at about 60°C.

The analysis of the kinetics indicated that the reaction may be viewed as occurring in two parts: an initial reaction, whose rate decays rapidly, and a steady state reaction, whose rate is constant with time. This is illustrated in Fig. 4.9, and may be expressed readily with Eq. 4.1.

$$P = \frac{K_I}{K_d} C_I^0 [M](1 - e^{-K_d t}) + K_{SS} C_{SS} [M] t \qquad (4.1)$$

Where $P =$ Total quantity of polymer produced at time t
   $K_I =$ Reaction rate constant for the initial reaction
   $K_d =$ Decay rate constant for the initial reaction
   $C_I^0 =$ Starting concentration of active centers for initial reaction
   $[M] =$ Monomer concentration
   $t =$ Reaction time
   $K_{SS} =$ Reaction rate constant for the steady state reaction
   $C_{SS} =$ Concentration of active centers for steady state reaction.

In this conceptualization, two types of active centers must exist simultaneously under the same conditions: one that retains its activity with time, and another that loses activity. The specific reasons for the different behaviors of the two types of sites were not clear.

Effect of TEA on Polymerization Rate. The initial polymerization rate, in the absence of any donor, increased with TEA concentration, at low TEA levels. This was generally accepted to be a result of (a) alkylating the potential catalyst sites, providing a metal-carbon bond, and (b) reducing the Ti^{+4} valence to Ti^{+3}. However, experimental results disagreed about the validity of a correlation between the Ti^{+3} concentration and activity. Regardless of that disagreement, the Ti^{+3} valence state was accepted to be a principal factor in propylene polymerization activity.

At higher TEA concentrations, although the initial polymerization rate increased further, the rate of decay became sufficiently rapid to greatly reduce the overall rate. This was explained various ways, including poisoning the active sites by replacing monomer with TEA, and over-reducing the Ti^{+3} to Ti^{+2}, inactive with propylene.

Effect on Isotacticity. In the above analysis of the reaction kinetics, the initial reaction (K_I) was determined to produce a highly isotactic polymer, about 90%, and the steady state reaction (K_{SS}) gave a less isotactic polymer, about 69% IsoIndex. Consequently, the IsoIndex decreased with polymerization time, and the IsoIndex relationship with key variables followed a consistent pattern over a range of conditions [100]. However, one would expect that pattern to be affected by changes in TEA or donor concentrations.

The effect of TEA on IsoIndex depended sharply on the presence and type of donor. Analysis of the IsoIndex behavior in the polymerization involved determining the productivity of isotactic polymer and atactic polymer separately by analyzing polymer samples taken during the reaction (for IsoIndex–time) or the final product (for effects of other variables).

No Donor. In the absence of any donor, increasing TEA concentration caused an increase in IsoIndex from a low level, about 30%, to about 50%. At the low TEA concentration, the IsoIndex would decrease with time, consistent with the kinetic analysis above.

Internal Donor. With an internal donor, IsoIndex was high, about 80%, at low TEA concentration, and approached 50% with higher TEA concentration [100]. The mechanism appeared to be:

1. The TEA initially replaced EB, primarily on isospecific sites, and activated those sites
2. The EB preferentially remained on, and maintained inactive, non-specific sites, resulting in low atactic polymer productivity, thus giving high IsoIndex
3. At higher concentration, TEA replaced EB on non-specific sites also, and IsoIndex approached the same equilibrium level reached with no donor, about 50%.

The above results correlated with chemical analysis of the catalysts, which showed that, as TEA concentration increased, EB concentrations dropped, and aluminum concentration increased equally. Titanium concentration remained essentially constant.

External Donor. In discussing the component of the catalyst using TEA plus external donor, we must consider the interactions between those two ingredients before examining the effect of the mixture on polymerization. The reactions of the alkyl and donor are understated as being "not simple" in our Barbe' reference [98].

As indicated by the change in the stretching frequency of the MPT C=O IR band, an acid-base complex forms, whose structure is not clear. There was even disagreement regarding whether one or two moles of the aluminum alkyl complex with the MPT molecule. Whatever occurred, it also rearranged, decomposed, and underwent secondary reactions, all of which were much discussed and little resolved in the literature. The effects of these poorly defined reaction products on the polymerization behavior remain largely uncertain, even today.

What was clear was that prolonged contact between the aluminum alkyl and the external donor caused the formation of a complex that was essentially inert, and therefore should be avoided. The following remarks assume a fresh mixture of MPT and TEA.

As illustrated in Fig. 4.10, up to about 0.2 molar ratio of MPT/TEA, atactic polymer productivity dropped sharply, while isotactic polymer productivity remained relatively constant, resulting in high IsoIndex, about 75%. In that phase, the initial reaction rate (K_I) dropped, and the steady state reaction rate (K_{SS}) was essentially unchanged.

As the MPT/TEA ratio increased over 0.2, atactic polymer productivity continued to drop substantially, while isotactic polymer productivity dropped more slowly, so IsoIndex increased further, in a "second step" of improvement, to about 95%. In this "second step," the steady-state reaction rate (K_{SS}) decreased.

The mechanisms were similar to those with internal donor only, in that the donor and TEA competed for occupation of the active sites, but the donor preferentially poisoned the non-specific sites.

Internal and External Donors. The behavior of the isotacticity with both internal and external donors was dramatically changed, from both the practical and mechanistic points of view. First, the overall polymerization rate was about 50% higher than that of an external or internal donor alone. Second, and most startling, that rate increase arose

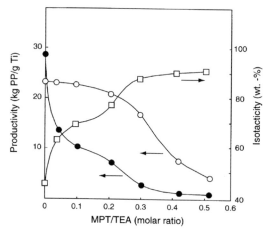

Figure 4.10   External donor effects on Isoindex. –□– Isotactic Index; –O– Productivity of isotactic polymer; –●– Productivity of atactic polymer. (Courtesy of Springer-Verlag, Berlin)

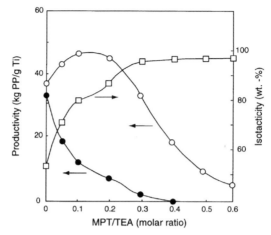

Figure 4.11 External and internal donor effects on Isoindex. —□— Isotactic Index; —O— Productivity of isotactic polymer; —●— Productivity of atactic polymer. (Courtesy of Springer-Verlag, Berlin)

from an increase in the productivity of the isotactic polymer with the initial increases in MPT concentration, as shown in Fig. 4.11, based on the Barbé reference. Accompanied by the usual decrease in atactic polymer productivity, the happy result was a significant boost in IsoIndex, accompanied by an increase in total polymerization rate. Further, the IsoIndex remained constant, or even increased with polymerization time.

It was many years later that Tait et al. suggested that the donor was converting the non-specific sites into isospecific sites, and increasing their reactivities as well [103]. Thus, a whole new phenomenon was occurring, and the reasons and mechanisms involved were not understood. However, the drive to understand the behavior, and then to exploit the understanding, continued.

4.5.3   Summary—Catalyst and Donor Chemistry

None of the theories on donor behavior explained everything observed. The theory of selective and reversible poisoning was consistent with formation of a complex between the Lewis base and $MgCl_2$, and explained well those cases where decreases in both atactic and isotactic polymer productivities were observed. However, these mechanisms fell short of explaining the changes in reaction kinetics, and were completely inadequate at explaining the increases in productivity of isotactic polymer.

5 Fourth Generation Supported PP Catalysts

5.1 Management Change at Montedison

From the point in 1976 when a commercially useful milled catalyst, FT-1, was available, and its chemistry was reasonably well understood, great changes began to occur at Montedison, at an ever-increasing pace.

In 1975, Dr. Italo Trapasso became responsible for Operations in Montedison, and immediately upon discovering the top management's neglect of research, began supporting the groups at the Ferrara and Novara locations. In addition to promoting the MPC agreement, he appointed Prof. Paolo Galli, with both technical and management credentials, to be director of the research center [84].

Professor Galli's management style was to praise, encourage, and inspire his people to seek, among other things, better understanding of the science. As a consequence, they were willing and able to do more, and to do it better. He knew and respected their technical capabilities, and they knew and respected his leadership. For the whole research center, this resulted in a renewed enthusiasm and rapid progress that carried them to heights none could have predicted at the time.

5.2 The Challenge: Success Through Understanding

To take advantage of the full potential of the supported catalyst, the morphology needed to be modified and controlled, in order to develop a new, more economical process. Prof. Galli, in his new role as director of the Ferrara research center, issued a memorandum in 1976 that renewed the call for a fundamental understanding of the catalyst behavior to reach the ultimate economic goals originally outlined in his 1970 report (See Section 3.2.3). He stated that while the FT-1 milled catalyst demonstrated that appreciable progress had taken place, the following goals remained [104]:

- Reach higher isotacticity, while maintaining activity
- Convert to a chemical source support to control particle morphology
- Ultimately, develop a spherical support, achievable from the alcoholate or hydrate, to realize the best process economics
- Apply all the advantages to a new process.

From this starting point, in parallel with the improvement and commercialization of the FT-1 catalyst, added attention was given to chemical source supports, spherical catalysts, and a solvent-free polymerization process, in order to reach the economic goals associated with those technical objectives.

5.3   Process Questions

5.3.1   Slurry Flow Behavior

Although milling generated the active form of $MgCl_2$, a large quantity of very fine particles was also produced. In addition to the variations in chemical nature and activity associated with the different particle sizes, the effects on processing were profound. The fines complicated the slurry process, whether diluent or bulk, in several ways:

- The slurry tended to "pack" and cause plugging in transfer lines and stirred tanks
- In order to avoid the plugging, low slurry densities were required, reducing the plant throughput
- Entrainment of the fines in liquid/solid separations dictated the use of extra separation stages or slower throughputs to cleanly separate the finest solids from the liquid phase.

Although some minor improvements could be effected by preparation and processing conditions, essentially all milled catalysts suffered from the same density and flow difficulties. Thus, the incentive to eliminate the high quantity of fines was great. The two major approaches, discussed in Sections 5.4 and 5.5, were (a) precipitated catalysts, with more uniform particle size than milled, and a granular shape, often called spheroidal in shape, and (b) controlled morphology[1] particles, where the catalyst particle support shape, size, and size distribution is determined by a physical process. Most often the physical form of the precipitated catalysts was defined by a chemical reaction that created a compound that was no longer soluble in the reaction medium. Only occasionally was the catalyst or precursor precipitated physically, by a change in temperature or solvent.

5.3.2   Isotacticity Targets

5.3.2.1   Process Differences

The IsoIndex level (in a homopolymer) that could be tolerated by a given process was very dependent upon the process itself. The atactic polymer was rather soluble in the diluents used in the slurry process, so even a small quantity would be dissolved by the

---

[1]Although the scientific literature often includes precipitated catalysts in the category "controlled morphology," I have chosen to limit that term to those catalysts coming from a process that provides direct and complete control over the catalyst shape and size, thus distinguishing them from precipitated catalysts.

diluent. While that dissolution reduced the stickiness of the solids in the slurry, it increased the diluent viscosity, inhibiting heat transfer, slurry flow, and solid/liquid separations. In addition, removal of the small amount of atactic polymer would require the usual diluent distillation, even at very high IsoIndex levels. Consequently, the total elimination of the atactic polymer removal step could not be expected at any feasible IsoIndex level with the slurry process. Only with a "total product" process, where all polymeric products are included in the solid PP product, could the atactic removal be dropped. It is here, in a total product process, that an IsoIndex level above about 95% was expected to result in an operable process. Thus, the bulk and gas phase processes received more attention.

In the bulk process, because the atactic polymer is substantially less soluble in liquid monomer than in diluent, a lower IsoIndex level could be tolerated without processing problems in the polymerization, where the liquid phase is still present; more of the atactic polymer remains in the polymer particles. After the flashing (vaporization) of the monomer, where the soluble polymer is deposited on, and to some degree, within the PP particles, stickiness would be encountered at some lower level of IsoIndex.

In the gas phase process, the atactic polymer is generated within the PP particles right from the beginning of the polymerization, so it is even less critical than in the bulk process. Thus, from the processability point of view, the IsoIndex level needed to eliminate the atactic polymer removal step decreased from impossibly high in the slurry, to moderate in bulk, to somewhat lower in the gas phase. As comparatively low IsoIndex polymers had been produced in the BASF gas phase process for years, there was little process reason to reduce the atactic content, but an improved product quality was highly desirable.

5.3.2.2   Copolymers

The earlier discussions regarding reaching about 95% IsoIndex to eliminate the atactic polymer removal step have been focused largely on a homopolymer product. However, the criteria are more severe if a copolymer is being considered. Both random and impact copolymers generate substantially more atactic material, and the catalyst must be still more stereospecific to deal with the increased atactic content. The impact (heterophasic) copolymer is worse than the random copolymer in this regard. Therefore, a company aiming at homopolymeric products would be satisfied with about 95% IsoIndex in a test polymerization. However, those seeking copolymers would require a higher IsoIndex level in the (homo)polymerization test to have comparable economics in the operation of a commercial copolymer process. The markets of interest to ME included injection-molded parts capable of surviving impact conditions in the northern European winter, requiring high impact copolymers, in addition to clear film markets, which called for highly substituted random copolymers. Consequently, the IsoIndex target was raised for ME, but commercial operations would be needed to define what precise level would be adequate. As we shall see later, Mitsui Petrochemical was less interested in the copolymer applications, and the strategies of these two companies began to diverge.

5.3.3 Particle Size

Meanwhile, the interest in copolymers at Montedison also generated a new process-related reason for using controlled morphology supports. The technology for providing low-temperature impact was well known: disperse up to about 40% of a rubbery phase component in PP, with good interfacial adhesion, at a particle size of about $0.5\,\mu$ [105]. This could be done, usually using EPR[1] as the rubber, either by extrusion compounding, or by generating the EPR within the PP particle in a copolymerization reactor downstream from the homopolymer stage. The more economical means of creating such an impact copolymer, also known as a heterophasic copolymer (sometimes incorrectly termed a block copolymer), is the reactor approach.

However, the presence of significant quantities of highly amorphous rubber in the polymer particle caused process problems. During the early stages of the copolymerization, the EPR filled any voids left by the homopolymer. Once the voids were filled, rubber began accumulating at the surface, and the particles in the polymerization environment became very sticky. Fouling of surfaces and lines occurred, and plugging usually resulted. In addition, the problem would be exaggerated when the second stage (rubber) reactor required the injection of additional TEA to maintain the reaction rate, as that would concentrate the rubber at the surface of the polymer particle [102]. The problem was to develop a reactor source impact copolymer that was processable. The FT-1 catalyst in the slurry process became sticky at low rubber levels, so it was not the answer. Consequently, both process and catalyst were being scrutinized for solutions.

Although the maximum processable rubber content was determined primarily by the void content, some benefit was expected by increasing the polymer particle size. Because the surface-to-volume ratio decreases with increasing size, it was thought that larger particles would tolerate a higher level of rubber at the surface before becoming sticky, and the momentum of the heavier particles in the flowing mixture were expected to be less likely to adhere to the reactor surface.

The particle size distribution would also be important in a gas phase process. The maximum particle size would determine the minimum conditions for fluidization of the solid bed, while the minimum particle size would determine the threshold of entrainment in the gas/solid separation step at the recirculating gas exit from the reactor.

Therefore, larger particle size and narrow PSD became added goals for a controlled morphology support to reduce the negative effect of higher rubber content, and to maintain processability in copolymerization processes. With these goals added to those already stated, interest in controlling the shape and size of the PP particle through the catalyst support was renewed with the Renaissance of research at Montedison in 1976.

---

[1]Although I have not differentiated between the "EPR" composition obtained from the rubber stage of the polymerization and the "EPR" purchased from rubber suppliers for extrusion compounding, these are recognized to be very different compositions. Without going into detail, the reactor "EPR" contains some crystallinity, compared to the essentially amorphous rubber from the rubber producers. For simplicity, the term "EPR" is used here for both materials.

5.4   Precipitated Catalysts

In order to produce the active crystalline form and have a more desirable morphology than is possible from milling, it is necessary to employ a chemical source of $MgCl_2$. As stated earlier, one can precipitate the support or generate the support in a particular geometry, usually consisting of a Mg salt or a complex of $MgCl_2$. To reach the double objectives of improved morphology and active $MgCl_2$, Montedison R&D management directed that the research effort be concentrated on chemical sources of $MgCl_2$, and that work on milled catalysts be reduced to a minimum. Initial work focused on precipitation.

Three schemes to a chemically based active $MgCl_2$ were under investigation at Ferrara by late 1976:

1. Formation of MgClOEt from Mg, $Si(OEt)_4$ and nBuCl,
2. Reaction of the MgClOEt with HCl to form $MgCl_2 \cdot 6EtOH$, and
3. Use of the Grignard reagent nBuMgCl, with HCl, EtOH, $SiCl_4$/EB, and $TiCl_4$, to form $MgCl_2 \cdot 6EtOH$.

The product from reaction scheme 1 could be converted into active $MgCl_2$ by treatment with $SiCl_4$ or $TiCl_4$. Treatment with HCl gave the ethanol complex $MgCl_2 \cdot 6EtOH$, as shown in scheme 2. The ethanol complex from any of the three schemes could be converted into active $MgCl_2$ by treatment with TEA, $SiCl_4$, or $TiCl_4$ [74]. Initial work used the TEA treatment.

Each of these schemes provided attractive levels of activity, with fair IsoIndex, as seen in Table 5.1, and all with much lower fines than milled catalysts.

However, in all cases, still higher levels of IsoIndex were needed.

5.4.1   Precipitated Catalyst Manufacturing Processes

The challenge to create a new chemical source support obviously included the need for a new process to manufacture that support, once it was defined. Work began at Montedison about 1976 on a simplified support process. An early trial catalyst, called FT-C, demonstrated that it was possible to form the $MgCl_2 \cdot n$EtOH complex with less than 6 molecules of EtOH simply by mixing the desired ratio of ingredients, then precipitating from solution. It was also shown possible to directly titanate the

Table 5.1   Performance of Chemical Source Supports

Source	Activity (kg PP/g Ti)	IsoIndex (%)
1. MgClOEt from Mg	290–350	90–91
2. $MgCl_2 \cdot n$EtOH from MgClOEt	305	89
3. $MgCl_2 \cdot n$EtOH from nBuMgCl	420	89

MgCl$_2$ · nEtOH complex without first removing the EtOH with TEA, thus achieving the titanation and formation of the active MgCl$_2$ simultaneously. The term "direct titanation" was used to describe this process step. This "direct titanation" catalyst performed better than those from the earlier processes. While the FT-C catalyst never reached commercial use, it was the predecessor to the simplified processes for the precipitated and spherical catalysts developed later.

5.4.1.1   GF2A

Beginning about 1976, a precipitated catalyst, GF2A, was prepared using schemes (1) and (2) above, and later employed the direct titanation procedure. It produced a granular support of narrow particle size distribution. A photo of typical GF2A polymer appears in Fig. 5.1. The early catalyst characteristics appear in Table 5.2.

To control the GF2A support particle size and distribution, an intense investigation was conducted in the lab and pilot plant to make it suitable for fiber grade polymer production. With proper processing, the average particle size and particle size distribution (PSD) could be produced in the same range as the old slurry "flake" particles. This particle size range served the substantial fiber business of Hercules and Moplefan, who had developed extensive technologies of mixing pigments, stabilizers, and other additives, with that physical form. Any major changes in the average particle size or PSD from the previously used "flake" would have required extensive changes in the mixing and extrusion procedures, and certainly would have upset the delicately balanced extrusion and fiber spinning operations. Because the GF2A polymer PSD was very similar to that of "flake," it was acceptable.

Because of matching the old "flake" PSD, GF2A was also suitable for obtaining a well-mixed peroxide additive to degrade the PP for fine filament non-woven operations.

Figure 5.1   Photo of PP from GF2A catalyst (25 ×)

Table 5.2   Typical GF2A Catalyst Composition and Performance (1978)

Analysis	Ti	2.5%
	Mg	20.5%
	Cl	67%
	EB	8.7%
	EtOH	1%
Polymerization[1]	Yield	240–280
	(kg PP/g Ti)	
	t-IsoIndex	94–95%
	bulk density	0.51–0.56
Polymer analysis	Ti	4 ppm
	Mg	32 ppm
	Cl	100–110 ppm
PSD of polymer	>1000 microns	0.5–1%
	>420	3–8%
	>177	89–84%
	>105	7%
	>53	0.4%
	<53	0.2%

[1]Conditions not reported; believed to be bulk, 70°C, 2 h.

The PSD of GF2A matched the "flake" well enough that dispersion of the peroxide was adequate, and the consequent molecular weight distribution was determined to be satisfactory by the producers of non-wovens, a group not given to rash experimentation.

5.4.1.2   TK

A more active catalyst, internally known as TK, was developed by MPC using a fourth scheme:

4. $MgCl_2$, dissolved in alcohol, was precipitated with $TiCl_4$ at a low temperature.

The TK catalyst gave uniform, fine particles and high activity, appropriate for a low cost process for homopolymer intended for applications such as fibers. The composition and performance of the TK catalyst is summarized in Table 5.3.

The photo in Fig. 5.2 illustrates the TK particles, which appear quite similar to those of GF2A, but are much finer.

The TK type, somewhat better than GF2A and early spherical catalyst (in the next section) with respect to low chlorine content, primarily due to its high activity, suffered from three shortcomings:

1.   The preparation was complex, and thus more expensive
2.   The particle size was difficult to control, and tended to be too fine, except for homopolymer production, and

Table 5.3   Typical TK Catalyst Composition and Performance (1981)

Analysis	Ti	3.5–4.5%
	Mg	20%
	Cl	61%
	EB	15%
Polymerization[1]	Yield (kg PP/g Ti)	300–380
	t-IsoIndex	95–97%
	bulk density	0.42–0.54
Polymer analysis	Ti	3 ppm
	Mg	15 ppm
	Cl	60 ppm
PSD of polymer	>1000 microns	0.3%
	>500	0.2%
	>177	23.4%
	>105	71.7%
	>53	4.3%
	<53	0.1%

[1]Conditions not reported; believed to be bulk, 70°C, 2 h.

Figure 5.2   Photo of PP from TK catalyst (20 ×)

3. The difficulties encountered with slurry flow and plugging with heterophasic copolymers were severe; it was not possible to operate with this catalyst in heterophasic copolymer production above a rubber content of about 12%.

This latter point was a serious difficulty for ME, who wished to have polymer grades with low-temperature impact resistance. MPC, on the other hand, with a narrower marketing target, not requiring the same level of low-temperature impact, was content with a highly active catalyst, suitable for homopolymer and low rubber copolymers. MPC proceeded with the manufacture and use of TK catalyst in their products, aimed primarily at fiber and film applications, and continued to use it for many years.

Thus, the precipitated catalysts reached a fairly high level of sophistication [106–111], while work was proceeding on the controlled morphology catalysts.

5.5   Controlled Morphology (Spherical) Supports

5.5.1   Spherical Support Process—Novara

In accordance with Prof. Galli's 1976 directive (See Section 5.2), the Novara lab was requested to develop a spherical support process based on the molten alcohol or H_2O complexes with $MgCl_2$ demonstrated on a lab scale for PE in 1970. Soon the Novara chemists had a process for making spherical support that employed spray cooling of molten $MgCl_2 \cdot 6EtOH$, and removing the alcohol with TEA treatment. Although that process provided a means of making a spherical support of adequate activity, the process was somewhat cumbersome and expensive. Further, the morphology of the support was inadequate. Work began at Ferrara about 1977 on refining the process.

5.5.2   Improved Spherical Support Process—Ferrara

As indicated in Section 5.4.1, it was determined with the FT-C catalyst that the $MgCl_2 \cdot 3EtOH$ complex could be made directly simply by mixing the proper ratio of $MgCl_2$ and EtOH. After precipitation of the complex, direct titanation was efficient and effective for creating the active form of $MgCl_2$, and was applied to the spherical catalyst program.

In an innovative change in approach, a whole new technology was implemented to control particle size. At Ferrara, a review of alternate approaches revealed that the size and shape of the support particles could be more readily controlled by emulsifying the molten phase in oil, followed by rapid cooling, or quenching, to solidify and crystallize the $MgCl_2 \cdot 3EtOH$ complex. The result, even in the early trials, was a higher fraction of spherical particles, with better shape, and a narrower size distribution. A lab-scale process for this improved spherical support process was developed and operating by 1981.

Composition and performance of the early Ferrara process spherical catalyst appears in Table 5.4.

Table 5.4 Typical Spherical Catalyst Composition and Performance (1978)

Analysis	Ti	3–3.5%
	Mg	17–19%
	Cl	59–61%
	EB	7–12%
Polymerization[1]	Yield (kg PP/g Ti)	250–350
	t-IsoIndex	93–95%
	bulk density	0.42–0.46
Polymer analysis	Ti	3–4 ppm
	Mg	17–23 ppm
	Cl	55–75 ppm
PSD of polymer	>1000	0.2%
	> 500	22.1%
	>177	59.5%
	>105	12.5%
	>53	3.0%
	<53	2.7%

[1]Conditions: bulk, 70°C, 2 h.

The spherical support and catalyst manufacturing process was optimized quickly, and major trials took place in pilot plant scale polymerization. The shape and size distribution of particles obtained from the spherical support process in 1982 was much better than the early versions, partly due to the addition of a final screening step. Fig. 5.3 shows the particle size distribution of typical spherical support, both as manufactured and screened, from the Ferrara process. Patent protection was obtained on this unique process [112, 113].

5.5.3 Morphological Control in Polymerization

Having developed a viable manufacturing process for spherical supports, the next challenge was to maintain that morphology in the PP manufacturing process, especially during polymerization. As mentioned in Section 3.3.1, there was a potential for particle disintegration during polymerization, which caused trouble in the Brindisi PE trial in 1971. Such problems, more frequently encountered with the more active catalysts, occurred at the very beginning of polymerization. One solution employed with the second generation catalysts was to conduct a small amount of polymerization in a separate step, usually 2–5 g PP/g catalyst, always under substantially more gentle conditions than in polymerization. Called "encapsulation" by Hercules and "prepolymerization" by ME and others, this was a widely known approach to retaining good particle morphology.

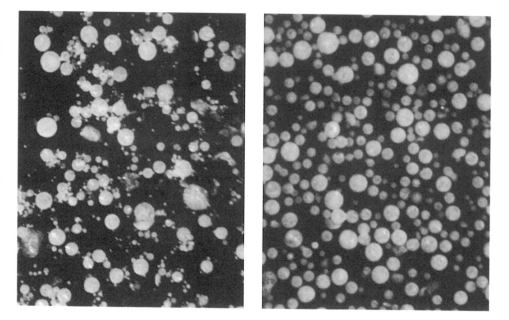

Figure 5.3   Photo of spherical support particles (60 ×): (a) raw product; (b) screened

It quickly became evident that in order to maintain the good spherical morphology of the new spherical support in the polymer, particle disintegration would have to be avoided, so prepolymerization was employed. Replication of the spherical support shape in the polymer particle occurred when prepolymerization took place under milder conditions prior to the first polymerization stage (See Section 5.10.1 for technical details). The use of prepolymerization represented another step forward in morphology control.

5.6   Donors—A Second Look

Now, let us return to the situation in 1978 following the run at Brindisi, when odor was recognized as a serious problem, and several approaches were taken to help solve this difficulty.

First, process changes were tried, in an attempt to remove the odor-causing components from the product. Steaming of the final product received the most emphasis. Although this was partially effective, it did not provide the improvement required. Thus, the attention turned to a change in donor.

When the odor difficulty occurred, the donor chemists were equal to the challenge. It became clear rather quickly that the aromatic group in the donor MPT was the principal cause of the odor. The group soon synthesized compounds that functioned similarly, but with less odor, such as *p-tert* butyl ethyl benzoate (PTBE).

Although PTBE was not quite as active as MPT, other catalyst improvements made it possible to maintain the overall productivity. A successful plant trial of this donor took place at Brindisi in late 1980, and work was begun on more economical synthesis of the donor. However, those results were quickly overtaken by the events described below.

5.6.1   The Silane Donors

The catalyst chemists had been studying the donor behavior through the mid-1970s, and they had already established the chemical requirements they felt would overcome many of the earlier shortcomings. Specifically, they sought:

1. Internal donors that would complex more strongly to $MgCl_2$, and
2. External donors that would;
   (a) form complexes with activated $MgCl_2$ that were more stable than those with TEA, and
   (b) be less reactive with TEA.

As part of the response to the odor problem, additional attention was given to a better understanding of donor behavior. Based on molecular modeling studies, the Novara group was confident that a bifunctional electron donor of the proper stereochemistry should form strong chelating complexes with the Mg atoms at the fracture surfaces.

Initial trials with bifunctional donors (DIBP, di*iso*butylphthalate, for example) and no external donor were disappointing in that about 80% of the DIBP was removed from the catalysts during polymerization due to nucleophilic attack by excess TEA on the Al-C bond to the carbonyl group coordinated to Mg. In comparison, only 50% of MPT was lost under the same conditions. However, the IsoIndex was 72% with DIBP compared to 57% with MPT, and exhibited a narrower MWD than with the mono-functional esters [114]. Based on the encouraging IsoIndex, narrow MWD, and theoretical considerations, the chemists were confident that the phthalates had the right structure for an effective internal donor, in spite of the above loss of DIBP. Consequently, phthalates were included in the continuing investigation into better donors.

Alkoxysilanes had been evaluated earlier as internal donors, and did not reveal any outstanding behavior. However, they had the advantage of being much less reactive with TEA, satisfying goal 2(b) above, so were tried again as external donors. In one test, after an unimpressive result, the mixture of alkoxysilane with TEA, believed to be non-reacting, was stored in the laboratory for several weeks. When the aged mixture was again used as an external donor, with phthalates as the internal donor, substantial improvements in both activity and IsoIndex were observed. In reviewing this happy result, it was eventually discovered that alkylalkoxysilanes, the products of a slow reaction between TEA and alkoxysilanes, had formed during the storage period, and were the active species that raised the IsoIndex over 95%, while also increasing the activity. Once again, a whole new level of performance had been achieved by the donor chemists, using a healthy mixture of systematic investigation and luck [73, 115].

More important, this high activity and IsoIndex performance held up well even at high Al/external donor ratios, the polymerization rate decayed less during the course of the polymerization, and the IsoIndex remained constant during the polymerization and over the full range of melt flow rates. The effectiveness of this donor combination was clearly outstanding, as shown in Table 5.5 and Fig. 5.4. Although neither was

Table 5.5   Silanes as 3rd Generation Catalyst Donors (Milled and Spherical/Slurry, 1980)

Catalyst	Activity, kg PP/g Ti	IsoIndex, %	Other Attributes
2nd Generation	12	95	
3rd Gen.; no donor	35	45	
3rd Gen.; internal donor	10	65	
3rd Gen.; external donor	30	94	
3rd Gen.; separate titanation	260	95.5	
3rd Gen.; silane A	350	98.0	OK at high Al/ED ratios; Low decay rate; constant IsoIndex vs time and MFR

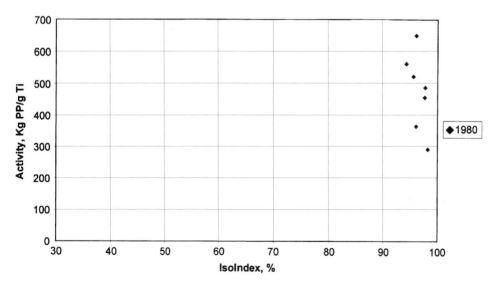

Figure 5.4   Performance of PP with Silane A, in slurry, milled support

particularly effective by itself, the combination of DIBP and an early alkylalkoxysilane, which we shall call Silane A, provided such outstanding performance that this is regarded as the stepping stone to the fourth generation of catalysts, and became the basis for a new round of patents [116–119].

These exceptional results have been attributed to the provision of the two aspects originally sought in the ED: strong complexing with the MgCl$_2$, and lack of reaction with TEA [120].

These results offered far better control of the reaction system, with a wider possible range of polymers. Further, the silane external donors were essentially odorless. Thus, the phthalate/silane combination quickly became the donors of choice, superseding the odorless donors mentioned above.

5.7   Implementation of Silane Donors

The effectiveness of the new phthalate/silane donor system was such that milled, precipitated, and spherical catalysts were all promptly converted to the silanes.

Table 5.6 shows the composition and performance of the spherical catalyst employing the silane donor system.

In the spherical catalyst, because of the 350% higher activity of the silane donors (Compare Tables 5.4 and 5.6), the diameter of the polymer particle could be increased by about 65% at the same catalyst size. By combining that improvement with a larger support particle size, the larger polymer particle size desired for the elimination of pelletizing extrusion and greater tolerance to high rubber content was suddenly achievable.

5.7.1   Impact Copolymers

The pressure to develop an effective impact copolymer (with about 40% rubber), that could be produced in the anticipated new PP process, using the new spherical support, was growing. There was considerable disagreement about what composition was needed. Many experts, both within and outside of ME, claimed that some type of graft copolymer was needed to develop sufficient adhesion between the homopolymer matrix and the dispersed rubber phase to reach the impact levels desired. Another group believed that the right rubber composition and particle size were the only requirements for adequate impact. This debate was not well documented, but Ferrara committed to the latter approach [121]. This position was supported by extrusion compounding experience, and, once proven true, it obviated the need for the additional graft copolymer reactor in the PP process.

With the third generation spherical catalyst (without silane), when trying to reach the high rubber levels needed for the impact requirements, the particles became sticky, and the process became inoperable. Discussions regarding the causes and prevention of stickiness were frequent, but not very productive; that would change a few years later.

Table 5.6   Typical Spherical/Silane Catalyst Composition and Performance (1982)

Analysis	Ti	2.6–3.0%
	Mg	19%
	Cl	64%
	DIBP	6–8%
	DIBP reaction products	0.5–1%
Polymerization[1]	Yield (kg PP/g Ti)	1200–1500
	t-IsoIndex	96.5–98%
	bulk density	0.39–0.45
Polymer analysis	Ti	1–1.5 ppm
	Mg	5–6 ppm
	Cl	20 ppm
PSD of polymer	>3000 microns	3.5%
	>2000	66%
	>1000	28%
	>500	2%
	<500	0.2%

[1]Conditions: bulk, 70°C, 2 h.

In addition, prior to the phthalate/silane donors, a major difficulty with preparing the high-rubber impact copolymers in the reactor was the loss of reactivity with time. This required the addition of more aluminum alkyl at the later stages of the reaction in order to revitalize the reaction rate. Unfortunately, that resulted in excess reaction at the surface of the particle, where the fresh TEA would reside, and it altered the Al/Ti ratio, with rapid development of a sticky surface. This seriously limited the ability to polymerize a sufficient amount of rubber in the copolymer stage to reach attractive impact values.

Considering that one of the outstanding benefits of the phthalate/silane spherical catalysts was the retention of a high polymerization rate even at long reaction times, it was possible to continue the reaction longer in the second stage without the injection of additional TEA. By also using a gas phase second stage reactor to eliminate the dissolution of the rubber, it was possible to reach rubber levels over 20%, with improved impact behavior. By having the reaction continue on the same catalyst centers used in the first stage, the rubber grew from within the particle rather than developing on the surface, and practical impact resistances were reached under operating conditions that were acceptable. This advance became part of the Spheripol process technology package, described in Section 6.1 [122].

While not fully appreciated at the time, the modestly higher porosity of the spherical catalyst (compared to milled or precipitated catalysts) was contributing to the ability to reach higher rubber levels.

5.8   An Explosion in Catalyst Research

5.8.1   General

Although the central theme of this book is that Montedison, with partner Mitsui Petrochemical, led the world in introducing major new catalyst technologies since 1968, other companies in the PP community also acted quickly and powerfully with their own contributions to the catalyst science. The following review, based on the open literature, while not extensive, reveals the intensity and diversity of those contributions.

Milled supports are not covered in detail, as precipitated or controlled morphology catalysts soon supplanted them. An inspection of the patents reveals that many avoided any mention of active $MgCl_2$. However, the *in situ* formation of active $MgCl_2$ has been more widely accepted than is evident from the patents, subsequent to the resolution of many patent disputes on the subject. Thus, active $MgCl_2$, although not always mentioned, is a constantly recurring theme in the catalyst developments described below.

5.8.2   Amoco

Amoco has been a serious participant in catalyst and process developments throughout the growth of PP. The parent, Standard Oil of Indiana, participated in the battle for the patent on crystalline PP in the US [123], and was awarded the basic patent in Canada. Immediately following the 1976 ME/MPC disclosure [86, 87], Amoco developed a variety of treatment procedures, pretreatments, extractions, and compositions that defined a proprietary region in milled Mg-based catalysts. The prevalent precipitation approach used the formation of Mg carbonates or carboxylates from the reaction of Mg alkoxide and CO_2, followed by the usual reaction with $TiCl_4$ and internal donor. Significantly lower fines were formed than in milled catalysts, the normal result for precipitated catalysts. A wide range of external donor structures has been covered [124–129].

Amoco Chemical also developed a precipitated catalyst beginning with complexed magnesium alkyls for forming the support. These catalysts and the process for their preparation have been described in detail [130].

Amoco has worked jointly with Chisso for many years in both process and catalyst areas. The Amoco catalysts are manufactured by Catalyst Resources Industries [131], and are employed in Amoco's horizontal stirred bed gas phase process. They probably represent a combination of the Amoco and Chisso technologies.

5.8.3   BASF, ICI (Targor)

The initial BASF approach to supported catalyst was the usual milled catalyst. The strategy soon turned to better morphology, forming a precipitate from Mg $(OR)_2$ and $TiCl_4$, with organic chloride and internal donors. After 1982, when the Montedison silane donors were revealed, BASF supplemented them with may other chemical effects, such as use of alcohol, extraction with hydrocarbons, or multiple TiCl$_4$ extractions.

In the later 1980s, the focus turned to inert carriers like silica or silica/alumina, the new aim being more controlled polymer particle morphology [132–144]. BASF and ICI formed a development agreement, and shared many results in both catalyst and process developments. In 1994, BASF acquired ICI's PP interests.

Prior to being purchased by BASF, ICI developed high yield supported catalysts using MgCl$_2$ or MgCl$_2$ on SiO$_2$, and pursued morphology control through spray drying, sometimes with dissolved polystyrene, as part of their cooperation in process development with BASF [145–148].

In 1997, BASF joined Hoechst to form Targor GmbH.

5.8.4   Mitsubishi Petrochemical and Chemical

Mitsubishi developed special treatments of Solvay catalysts prior to the 1976 breakthrough by ME/MPC, and many supported catalysts thereafter. They developed some highly active precipitated catalysts in the early 1980s, using MgCl$_2 \cdot$ 2Ti(OBu)$_4$ with methylhydropolysiloxane, and effecting precipitation by reduction to trivalent titanium. They also worked on SiO$_2$ support, but focused on donor behavior. Besides specific silane donors and special procedures, they developed a catalyst not requiring an external donor, and another not using an organic acid ester, thus having reduced odor [149–159]. Mitsubushi was continuously active in catalyst research, discovered numerous unusual chemical aspects of catalytic behavior, and frequently reported their results in technical reviews.

5.8.5   Mitsui Petrochemical

Because MPC and Montedison shared developments after 1975, their catalyst technologies are very similar. ME patents are normally used to illustrate the technical points in this book, while typical patents from the MPC technology are included here for reference [110, 111, 160–166].

5.8.6   Phillips

Phillips, who remained very strong in the silica/alumina supported chromium oxide catalysts for PE, and who won the US patent rights to crystalline PP [123], continued to develop titanium-based catalysts as well. Following work on second generation catalysts, early work on supported catalysts focused on milled forms, then switched to precipitated Mg compounds or complexes, with a wide range of chemical approaches, including treatment with phosphites, halohydrocarbons, or heat, in hydrocarbon solvents. In some instances, precipitation of the soluble hydrated or alcoholated

MgCl$_2$ complex was effected with an aluminum alkyl chloride. Recent work has aimed more at directly controlling the catalyst morphology [167–170].

5.8.7   Shell[1]

When the ME/MPC patent appeared in 1976, Shell began more serious work on supported catalysts at the Amsterdam research center (Shell International Research). They soon joined in the parade of milled MgCl$_2$ supported catalysts [171].

Their work then led to the reaction of magnesium ethoxides with TiCl$_4$, with donors and halohydrocarbons present, to precipitate MgCl$_2$ *in situ*, and giving catalysts with very high activity and selectivity [172]. In 1981, however, this work was interrupted when a business decision was made to suspend research in Amsterdam. Shell Chemical Company at Houston continued development of the high activity catalyst. This led to the introduction of the catalyst into a plant a Norco LA in 1984. In 1983, a reversal of the earlier management decision led to resumption of research in Amsterdam, and by the end of the 1980s, high activity catalysts were being used commercially in various plants using the Shell LIPP (liquid propylene polymerization) process.

The first developments involved catalysts containing ethyl benzoate (EB) as internal donor, and were later extended to diesters, following the disclosure of phthalate ester/silane combinations in the Montedison patents issued in 1982. This results in catalysts having lower initial activity and a slower decay rate. Similar improvements were achieved by an alternative approach involving catalysts of the type MgCl$_2$/TiCl$_3$OAr/EB, in which the activity/decay profile of the catalyst could be controlled by proper choice of the Ti component; the (OAr) group represents an aryloxide. This catalyst could be prepared via a simple one-step reaction of Mg(OAr)$_2$ with TiCl$_4$. Moreover, spheroidal morphology was achieved by controlled precipitation of the Mg(OAr)$_2$ in the reaction of Mg(OEt)$_2$ with the corresponding phenol. Earlier approaches to morphology control had involved the preparation of spheroidal Mg alkoxides and alkyl carbonates via spray drying [173–180]. The Shell developments did not emphasize the high impact copolymers.

By the mid-1980s, Shell Chemical Company in the USA was working in close cooperation with Union Carbide, providing much of the catalyst know-how for their joint venture for Unipol Process PP, using Union Carbide's gas phase process. Shell International Research was not part of that joint venture. The first commercial Unipol Process PP facility began production in 1985. In 1995 Shell's worldwide interests in PP outside of the US became part of Montell Polyolefins. Shell's catalyst and PP manufacturing facilities in the US were sold to Union Carbide in 1996.

---

[1]Although the PP assets from Shell (Excluding the US) were joined with Himont to form Montell in 1995, most of the developments described in this book occurred before the Montell joint venture. For this reason, the Shell catalyst developments are listed separately.

5.8.8   Toho Titanium

Toho has developed technology using $Mg(OEt)_2$, $Mg(OEt)Cl$, and Mg carboxylates, the last to give precipitated catalysts, in aromatic solvent, similar to the technology of Shell. They have often included magnesium stearate or laurate in their catalyst patents. Toho manufactures catalysts, and is a strong competitor in catalyst sales, particularly in the Pacific rim area [181–186].

5.8.9   Union Carbide

Union Carbide (UCC) has been very active for many years in both catalyst and process developments for polyethylene, utilizing the gas phase Unipol Process. As part of the joint venture with Shell Chemical Company in the US for Unipol Process PP, UCC has received substantial catalyst know-how from Shell Chemical, and has also conducted considerable work in PP catalyst development since the start of that joint venture [187].

At the formation of Montell in 1995, UCC retained rights to the Shell catalyst technology for use both internally and for Unipol Process licensees. Thus, UCC is a robust participant in PP production, process licensing, and catalyst sales. Much of the UCC PE catalyst work used SiO_2 support, but it was not sufficiently effective for PP.

5.8.10   Other Companies

The above companies represent the major ones active in the development and commercialization of supported catalysts for PP. They either licensed their catalyst know-how and sold their catalysts, or packaged their catalyst know-how as part of a process technology package. However, many of the following companies have also established proprietary regions of technology in supported catalysts.

- *Chisso Corp.* Chisso worked out techniques for obtaining good morphology with the $TiCl_3$ type catalysts. Their efforts in supported catalyst also focused on morphology. Use of $Mg(OR)_2$ plus CO_2 to develop soluble complexes, followed by precipitation with a chlorinating compound, like $TiCl_4$, or spray cooling the $MgCl_2 \cdot$ alcohol complex to get largely spherical particles were developed. Chisso has worked with Amoco in catalysts, and both technologies are probably used for the catalysts in the Amoco/Chisso gas phase process [188–191].
- *Akzo, Stauffer, and Toyo Stauffer Chemical Co.* This group created precipitated catalysts based on MgR_2, alkoxy Mg halides, or Grignard reagents, with CCl_4, acid ester donor, and phenol, followed by $TiCl_4$, with a very good activity and IsoIndex [128, 192–195]. Akzo bought the former Stauffer facilities, and has been the manufacturer of Montell and Mitsui supported catalyst components in the Americas.
- *Exxon.* Exxon has focused on sterically hindered cocatalyst components with major advantages in stereochemistry of polymers [196, 197], but have not been a major factor in supported catalyst chemistry or licensing. However, a new catalyst was

announced with the startup of their plant in Bayport in 1981 [198]. Their main effort has been in metallocene catalysts since the mid-1980s, and a strong position exists there.

- *Hoechst (Targor)*. Hoechst was very active in early catalysts, using the $Mg(OR)_2$ plus $TiX_n(OR)_{4-n}$ approach similar to that of Shell. Like Exxon, they have concentrated on metallocenes more recently [199–206]. Hoechst joined ICI in 1997 to form Targor GmbH.
- *Neste Oy**. The Borealis/Neste Oy technology uses $MgCl_2$ from various Mg compounds, complexed with alcohol, and focuses on spherical morphology and controlled porosity, similar to the Montell technology [207–210].
- *Sumitomo*. Sumitomo developed precipitated catalysts with good activity, especially high IsoIndex, good morphology, and high bulk density based on Grignard reagents, aluminum cresoxides and phenoxides, titanium alkoxides, ethers, phenol, and organic acid esters. A range of highly hindered silanes and piperidines were covered as external donors [211–215]. Sumitomo manufactures and uses their own catalysts, and licenses their catalyst technology.
- *Asahi, Denki, Idemitsu, Nissan, Showa Denko, Toso, and Tonen*. There has also been considerable activity in catalyst development by these companies.

5.9   Fourth Generation PP Catalysts

At Montedison in 1980, the combination of:

- The ability to control the catalyst morphology with the spherical support,
- The high activity and IsoIndex of the phthalate/silane catalyst system over a wide range of conditions and reaction times, and
- A viable scheme for good impact copolymers,

set the stage for a major new advance in the PP manufacturing process, with the economics anticipated in 1970. The contrast between the situation in 1975 when the neglect by the ME management was ended, and in 1980, when the results exceeded even the fondest hopes of the research and process groups at Ferrara and Novara, is a tribute to the people in those groups. It is still a wonder how such a small, modestly funded laboratory, with only rudimentary support, catalyst, and polymerization pilot plants, was able to accomplish so much in such a short time. All those individuals who participated in the successful development of these new catalyst and process technologies are to be commended for their outstanding achievements.

The first catalyst system to be considered entirely within the fourth generation was the combination of the phthalate/silane donors and the spherical support, using liquid

---

*Now Borealis, with Statoil.

Table 5.7   Silanes as 4th Generation Donors (Spherical Support/Liquid Monomer, 1981)

Catalyst	Activity, kg PP/g Ti	IsoIndex, %	Other attributes
Silane A	1870	96.3	Low decay rate; constant IsoIndex vs time and MFR

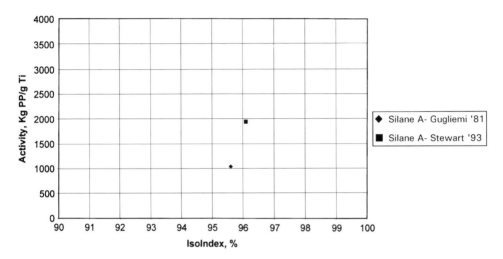

Figure 5.5   Performance of PP with Silane A, in liquid monomer, spherical support

monomer in the homopolymer reactor. The activity/IsoIndex behavior of this preferred system, using the DIBP/Silane A donors, is shown in Table 5.7 and Fig. 5.5. With the use of liquid monomer and the spherical support, the numbers have different specific meaning than the milled, slurry results, so this table represents the beginning of a new series of comparisons. However, while these results are not technically comparable to earlier tables, the numbers still reflect the progress being made in the ability to produce PP at a high rate with good IsoIndex levels.

5.10   Catalyst Concepts, Mid-1980s

With the arrival of the silane donors and spherical supports, the chemistry of the fourth generation catalyst was largely defined. In accordance with the 1976 Galli directive to reach better performance through better understanding, significant advances in the understanding of the technologies had been reached at ME and MPC by the mid-1980s. Those relating to the development of spherical support particles, while crucial to the success of the venture, were highly proprietary, and will not be discussed here. The

Spheripol process is adequately described in the literature [102, 216]. Below are more detailed discussions regarding replication and donor chemistry.

5.10.1 Replication of the Catalyst Particle Shape

One of the key factors fostering the success of the spherical catalyst was the understanding of the conditions that permit or prevent the replication of the catalyst particle shape. Galli had listed the requirements in general terms in his 1970 study. From more detailed understanding of this phenomenon, it was possible to quantify the conditions governing replication.

5.10.1.1 Process Limitations to Replication

The following definitions are used in this discussion, and throughout this document:

Disintegration: Breakup of the growing polymer particle into irregular shaped pieces due to uneven polymer growth, normally occurring during rapid growth early in the polymerization, usually caused by reactor concentration gradients arising from diffusion limitations. This commonly results from monomer gradients, but other reactants (cocatalyst, active catalyst sites) can be involved. Thermal gradients, arising from heat transfer limitations, can also cause disintegration. The term "fragmentation" has often been used in reference to this process.

Subdivision: Division of the catalyst particle into primary crystallites, embedded in the growing polymer particle, which ideally would replicate the catalyst shape. Some researchers (Ferrero, Choi) have used "fragmentation" in reference to this process.

Attrition: Break up of the catalyst or polymer particle into irregular shaped pieces due to processing, such as agitation or pumping.

The replication of the catalyst shape can only be achieved within certain sets of polymerization conditions. The greatest difficulties with disintegration of the catalyst particle, based on both the mathematical model and actual experience, occur at low diffusivities (high resistance to diffusion) and early in the polymerization, as indicated in Eq. 5.1 [102].

$$\frac{dy}{dt} = \frac{K \cdot C \cdot [M]}{1 + \frac{\tau_D}{\tau_R} \cdot y} \tag{5.1}$$

Where y = polymer yield (g PP/g Ti),
dy/dt = overall polymerization rate (g PP/(g Ti·s)),
K = reaction rate constant (g PP/(g active Ti·(mol/L)·s)),
C = concentration of active centers (g active Ti/g total Ti),
$[M]$ = monomer concentration (mol/L),
τ_D = characteristic time for diffusion (s), and
τ_R = characteristic time for reaction (s).

In this expression, when diffusion rate is very high relative to the reaction rate (reaction-limited), and, thus, τ_D is very small relative to τ_R, the denominator approaches 1.0, and the reaction is under kinetic control, as may be seen in Eq. 5.2.

$$\frac{dy}{dt} \cong K \cdot C \cdot [M].\tag{5.2}$$

All the active centers are consequently equally accessible to the monomer, and replication is favored.

In the instance when the reaction rate is very high relative to the diffusion rate (diffusion-limited), the second term in the denominator of Eq. 5.1 becomes large, and, since Eq. 5.3 is true, then Eq. 5.4 is operative, and the polymerization rate becomes inversely proportional to the yield.

$$\tau_R \times K \cdot C \cdot [M] = 1\tag{5.3}$$

$$\frac{dy}{dt} \cong \frac{1}{\tau_D \cdot y}.\tag{5.4}$$

Thus in the first moments of the reaction, the polymerization rate will be highest in a diffusion-limited reaction. In addition, the particle growth as a percentage of particle size is extremely high at that time, and any stresses due to uneven polymer growth rates will be exaggerated. Further, the particle surface area is smallest at the beginning, presenting the greatest resistance to diffusion of the monomer or cocatalyst into the particle. Consequently, the danger of disintegration is greatest at the beginning of the reaction.

In the instance of the diffusion-limited reaction, substantial monomer (or other reactant) concentration gradients can exist through the radius of the particle, with accompanying gradients in reactions rates [217, 218]. As a consequence, the outer layers can grow much faster than the inner layers. Without describing the specific physical consequences, Hutchinson et al [218] calculate a degree of voiding associated with this uneven growth. In the extreme situation, the outer, more reacted layer, will separate from the inner layers, and will disintegrate into fragments.

Various combinations of shells and highly voided interiors are possible. One of the behaviors observed in Galli's polymerization of PE in spherical supports in 1970 appears in Fig. 5.6, concentric shells, also known as the "onion" structure. Little has been written about such structures since, although much effort has been expended to avoid them; see "Prepolymerization," below.

Less pronounced, but similar "onion" structures have occasionally been seen in PP particles, such as that illustrated in Fig. 5.7. The conditions used to polymerize that particle, from a Ferrara pilot plant run, were normal, so the specific reasons for the formation of the "onion" structure are not known at the time of this writing. Similar onion-like structures have also been observed by the author in 1980 second-generation PP particles and in 1997 high-porosity particles. Although the general consensus has been that the onion structure does not occur in PP, due to its slower reaction rate than

Figure 5.6   "Onion" structure in PE (1970)

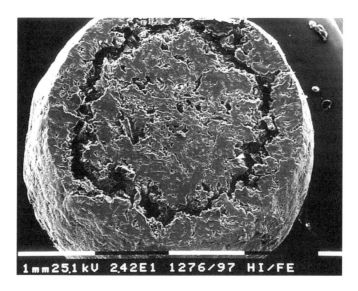

Figure 5.7   "Onion" structure in pilot plant PP (1996)

PE, the above observations suggest that the possibility may be more likely than previously believed. Further study of the factors causing these kinds of formations may be warranted.

5.10.1.2 Prepolymerization

In order to avoid the difficulties described above, the initial polymerization must take place in a separate stage, at very mild conditions; this is the concept of prepolymerization. By prepolymerizing a small amount of PP at low monomer concentration and/or low temperature, the critical initial phase can be passed without the particle disintegration that would result from normal polymerization conditions. In the early 1980s, extensive work was conducted at ME in the continuous pilot plant to establish conditions for contacting the catalyst components with each other (precontact) and prepolymerization, in order to maximize the performance of the spherical catalysts [219]. Thermal gradients can also contribute to the tendency for disintegration of the growing polymer particle. The dangers arising from thermal excursions are also calculated to diminish with prepolymerization [220]. Consequently, precontact and prepolymerization have been employed extensively to avoid disintegration in both milled and chemical source catalysts.

It was because of the above understanding of the criticality of the initial polymerization period that prepolymerization has been employed, particularly since the development of the high quality spherical support in 1982. A yield of about $100\,g$ PP/g catalyst is usually reached in the prepolymerization of fourth generation catalysts.

Besides preserving the ability to replicate the catalyst particle, prepolymerization also significantly increases the catalyst activity. The specific reasons for this behavior are not clear.

5.10.2 Donor Chemistry

The chemistry of the third generation monofunctional donor combinations was understood far better in 1983 than in 1976, as described in several publications [75, 114, 120]. The goal of the external donor was to keep the total donor concentration on the catalyst constant at about 10%. For the monofunctional donors, this required maintaining the TEA/MPT molar ratio low (about 3) to avoid secondary reactions between TEA and MPT. Further improvement in donor performance would require:

1. An internal donor with stronger coordination with $MgCl_2$
2. An external donor that would compete better with TEA for coordination with $MgCl_2$, and
3. No chemical reactions with TEA.

Molecular modeling calculations indicated that the fourth generation multifunctional donors, such as phthalate esters, with the right distance between donor atoms, should form stronger chelating complexes with the tetra coordinated Mg atoms on the

(110) plane or binuclear complexes with two penta coordinated Mg atoms on the (100) plane, the sites of aspecific polymerization [221]. Although, as predicted, di*iso*butyl phthalate (DIBP) as an internal donor provided higher IsoIndex than EB or MPT, the DIBP was rapidly removed from the catalysts by reaction with TEA.

Silanes, tested earlier, had displayed unexceptional performance as internal donors, but they did exhibit a high ability to compete with TEA for complexing with $MgCl_2$, and inertness toward TEA, two of the requirements mentioned above. Thus, when polyalkoxysilanes were used as external donors, as described in Section 5.6.1, exceptionally high stereospecificity was achieved, as anticipated, and unusually high activity, which was not expected, also occurred. Further, the inertness toward TEA resulted in the ability to reach the above effects even at high Al/donor ratios, as illustrated in Fig. 5.8 [74].

Although the phthalate esters react with TEA, and the silanes react with $TiCl_4$, the phthalate/silane system satisfied most of the requirements for the ideal donor:

1. Polyfunctionality
2. Distance between donor atoms of 2.5–3.3 Å
3. Coordination with $MgCl_2$ in the presence of TEA and $TiCl_4$
4. Absence of secondary reactions with $TiCl_4$ during catalyst synthesis, or Al-C, Ti-C, or Ti-H bonds during polymerization.

The fact that the substantial improvement in isotacticity provided by the phthalate/ silane donor system was also accompanied by an increase in activity was convincing evidence that the donor was acting as suspected earlier. It was converting the non-specific sites into isospecific ones by altering the environment of the active site,

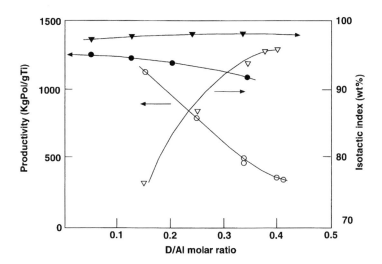

Figure 5.8   Comparison of 3rd and 4th generation catalysts: sensitivity to TEA concentration. ○, ▽ 3rd generation; ●, ▼ 4th generation. (Courtesy of Reed Elsevier, London)

inducing chirality to adjacent Ti atoms [75, 222, 223]. Because the isospecific sites exhibit $4 \times$ to $10 \times$ higher reaction rate than the non-specific sites, the overall productivity increased as well [224]. Several studies, of (a) molecular weight distribution and the precise steric makeup of the isotactic polymer fraction [225], (b) changes in the reaction kinetics such as lower reaction decay rate, (c) activity maximum at 80°C instead of 60°C, and (d) constant IsoIndex with reaction time [102], confirmed the view that the donor participates in the conversion of non-specific sites into isospecific sites.

Part III

The Revolution in the PP Industry

6   New Manufacturing Processes

The commercial consequences of the fourth generation catalysts were truly revolutionary. The catalyst chemists and the process engineers suddenly found their long-term objectives were in hand. The ideal catalysts they had aimed at for decades had become a fact of life. High activity, low decay rate, good IsoIndex, wide range of melt flow rates, impact copolymers, MWD control, and morphology control were all in place. The process engineers were no longer confined to the conditions dictated by the limited catalyst capabilities. Thus began the most obvious next stage of the rebirth of the PP industry: renewal of the manufacturing processes.

Once again, ME and MPC led the way, but many PP producers designed new versions of PP flowsheets based on their own process experiences and aimed at their own business needs [226]. There was a veritable process design frenzy in the 1980s. Many producers who formerly limited their efforts to homopolymers expanded, and were then able to make the more sophisticated random and impact copolymers.

Along with the revolution in processes came a wider range of properties in PP (Chapter 7), followed by major commercial shifts in the PP industry (Chapter 8).

6.1   Spheripol Process (Montedison/Himont)

Although the Spheripol process was designed and scaled up by Montedison, it was followed quickly by the formation of Himont, and is thus usually associated with Himont. It is currently operated and offered by Montell (See Section 8.1)

The principal benefit of the fourth generation catalysts was the freedom from the design limits that encumbered the old processes. The ME process group was instructed in 1979 to design a process based on first principles that took advantage of the outstanding catalyst activity, stereospecificity, and the promising, but as yet undemonstrated, spherical support. Once again, the new vigorous research management was planning for success.

Due to limited pilot plant facilities, there was much uncertainty about the approach to take regarding a new process design. While numerous advantages soon presented themselves as a result of the achievement of fourth generation catalysts, many of the process concepts being considered were untested. Consequently, numerous discussions took place involving the choice between a bold process design and a more conservative approach. In theory, the most aggressive design would take the best advantage of the catalyst capabilities, but lacked any demonstrated commercial performance, compared to known process equipment and procedures that were well understood in the industry, but would not permit the optimum use of the catalyst.

As these discussions proceeded, the process people at Montedison, led principally by Ing. G. DiDrusco, became increasingly convinced that they should continue to aim for the ultimate goal: the best process possible. With a minimum of pilot plant support information, they proceeded to design the Spheripol process, including numerous innovations:

- A liquid-filled reactor
- A monomer flash evaporation under pressure that allowed liquification of the monomer just by cooling, and recycling by pumping the liquid rather than using an energy consuming gas compressor
- Prepolymerization to a high level (50–100 g PP/g cat vs the 2–3 g/g usual) to present a tougher particle to the first reaction stage
- Continuous rather than batch precontact and prepolymerization that guaranteed a constant prepolymer quality, essential for the subsequent polymerization steps
- A liquid-phase first stage, and a gas-phase second stage reactor for the optimized operation of homopolymer in stage one, and EPR in stage two.

Although some of the details evolved further, the concept initially selected did not vary: aim for the best process. Considering the number of new process operations used, and the simplicity and reliability of the final design, with its consequent attractive

Table 6.1 Advantages of the Spheripol Process

Liquid monomer advantages

- High output per unit volume (>300 kg PP/h m^3)
- Small product changeover inventory
- Good heat exchange between liquid monomer and polymer particles
- Cocatalyst and donor: soluble, homogeneously distributed
- Low mixing energy requirements
- Full reactor; no fouling problems from evaporation into the gas space
- Monomer easily condensed for recycling

Loop reactor advantages, for homopolymer and random copolymer

- Good circulation easily achieved; high polymer concentration possible
- Good heat transfer
- Narrow diameter tube allows thin wall even at high pressure

Gas phase reactor advantages, for rubbery copolymer

- No dissolution of rubber in liquid monomer or solvent
- Good particle flowability; no sticking
- Cooling by external gas circulation
- Good temperature and composition uniformity in fluid bed
- Good fluidization and separation when PSD is narrow
- Reaction rate maintained with original catalyst

economics, the Spheripol process is a tribute to the abilities and creativity of the design engineers.

The principal advantages of the design used [102] are summarized in Table 6.1.

The Spheripol process appears in Fig. 6.1 [216], and became an important part of the Montedison technology. A decision was made to construct a new plant based on it, although the ME management was far from unanimous in that decision. The new plant was constructed at Brindisi, and initial operations took place in late 1982.

Even troubles with the Brindisi start-up were quickly converted into better understanding. At first, the polymerization simply did not proceed anything like the pilot plant operation; the catalyst performance was totally inadequate. The process of precontacting the external donor with the aluminum alkyl prior to mixing it with the catalyst was employed. However, the residence time in the transfer pipe in the commercial plant was significantly lower than in the pilot plant, because of the higher throughput in the commercial plant. After some study, it was realized that the aluminum alkyl/external donor complex required some time to form. When the precontact residence time was raised, operation returned to normal, and a better understanding of the donor complexation reaction was gained [227].

Following the above change, the operation of the Brindisi plant became a dramatic demonstration of the efficiency and economy of the high activity, high stereospecificity, spherical catalyst in the Spheripol process. Meanwhile, other PP producers developed

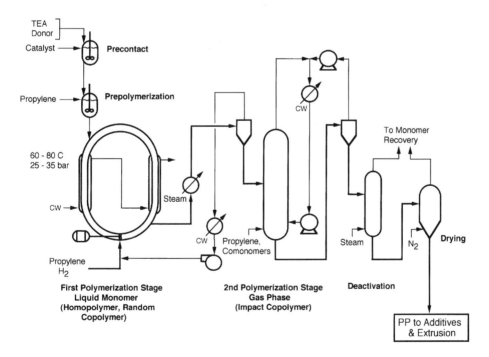

Figure 6.1   Spheripol process (Himont/Montell)

their own versions of the PP process that took advantage of the capabilities of the fourth generation catalysts.

6.2   Hypol Process (Mitsui Petrochemical)

Building on the same technology shared with ME, MPC elected a more conventional design that satisfied their market aims; their Hypol process is shown in Fig. 6.2. Combined with the more active, but less versatile TK catalyst, the Hypol process was well suited for the production of homopolymer at a low cost. A gas phase reactor of small size is used for reacting down the monomer in homopolymer production, or a larger one may be used for production of copolymers with moderate levels of impact.

6.3   Fluid Bed Gas Phase Unipol PP Process (Union Carbide)

Union Carbide had extensive experience with the fluid bed gas phase Unipol Process in polyethylene. Given a polypropylene catalyst with adequate activity, IsoIndex, and

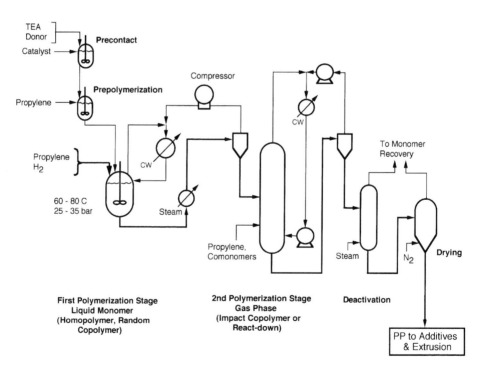

Figure 6.2   Hypol process (Mitsui Petrochemical)

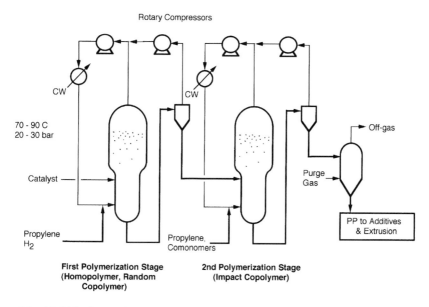

Figure 6.3   Fluid bed gas phase Unipol PP process (Union Carbide)

morphology, derived from their own and Shell's activities, the manufacture of PP in the Unipol Process has proven to be very effective. A demonstration plant was built at Seadrift, TX, USA, and operations began in 1985, the principal objective being licensing rather than production and sales. This gas phase process can be constructed in very large plants, affording excellent economics, and may be configured for materials ranging from homopolymer to impact copolymers. The flowsheet of the Unipol PP Process for impact copolymers appears in Fig. 6.3 [228–231]. The Unipol PP Process captured an appreciable portion of the worldwide licenses during the major expansions in the 1990s.

6.4   Vertical Stirred Gas Phase Process (BASF)

As stated earlier, the BASF gas phase process had no need of further simplification, but enjoyed improvements in product quality, due to the reduced catalyst residues and lower atactic polymer content. In addition, a wider range of melt flow rates and comonomers could be accommodated with the fourth generation catalysts without encountering operating problems.

By adding a second reactor, impact copolymer capability was achieved. The second stage was operated at somewhat lower pressure and temperature, requiring more compression to liquify the recirculated gas [26].

6.5   Bulk Process (Rexene, Phillips, Exxon, Sumitomo)

With the high activity, high IsoIndex catalysts, the bulk processes were able to drop the workup sections, and the chemical procedures for altering the solubility of the catalyst residues, simplifying the plants and thus improving the economics. The simpler Rexene (Dart) process appears in Fig. 6.4 [232], and the Phillips process in Fig. 6.5 [36].

Exxon started a plant in the late 1980s using a stirred reactor bulk process licensed from Sumitomo [233]. The flowsheet included the countercurrent washing of the PP/monomer slurry with fresh monomer using a column developed by Sumitomo. The washing operation provided polymers with low residues and low levels of oligomers. That refinement required a subsequent distillation step to separate the propylene from the solubles for recycle. The flowsheet appears in Fig. 6.6.

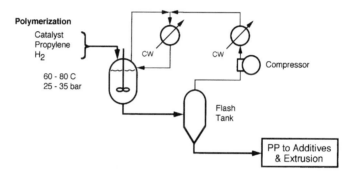

Figure 6.4   Bulk process (Rexene, 4th generation)

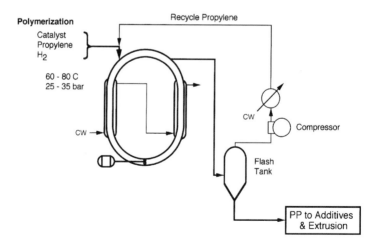

Figure 6.5   Bulk loop process (Phillips, 4th generation)

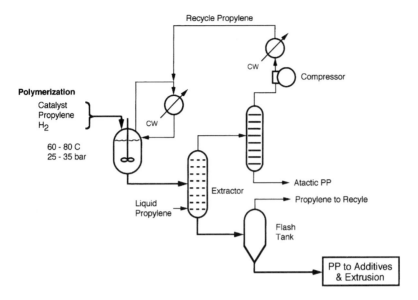

Figure 6.6   Bulk process (Exxon, Sumitomo)

6.6   Horizontal Stirred Gas Phase (Amoco, Chisso)

Amoco has been active in catalyst and process development, as well as production, from the earliest days of PP, and began their cooperation with Chisso in 1960. In 1975, gas phase process development was begun in earnest. The key feature is the horizontal stirred gas phase reactor, in which the polymer flows through baffled zones in a pattern approaching plug flow. The first commercial trial of the stirred horizontal reactor took place at Chocolate Bayou in 1979, using a second generation catalyst. The reactor details appear in Fig. 6.7. Condensed monomer is returned to the reactor for cooling [234, 235].

In addition to the zoned polymer flow, the gaseous monomer inlets, gas outlets, catalyst injection, and liquid monomer addition are zoned enough to allow some staging of the temperatures and gas compositions. Amoco does not employ prepolymerization, probably because the first zone can be operated at a reduced reaction rate.

Chisso began working on a heterophasic copolymer process in 1982 using their own high activity high stereospecificity controlled morphology third generation catalyst. The first commercial scale impact copolymer plant was operated in 1987 in Japan, and the second at Chocolate Bayou in 1992. The copolymer process employs two reactors arranged vertically to assist the polymer flow into the second reactor, as shown in Fig. 6.8.

The plug flow pattern narrows the residence time distribution, and thus provides a more uniform particle at the outlet. This would become more important in impact copolymers, where the frequency of the undesirable extremes, low residence time in the

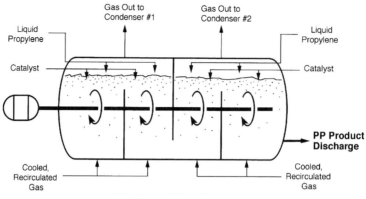

Note: Stirrer paddles not shown

Figure 6.7   Details of horizontal stirred reaction (Amoco)

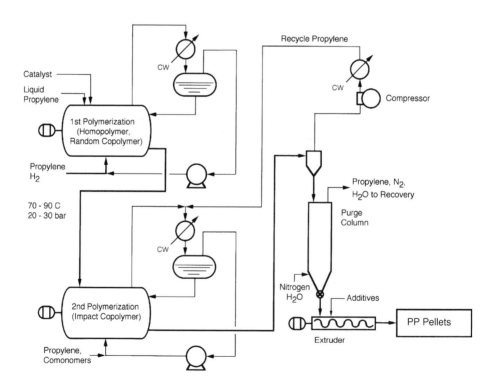

Figure 6.8   Horizontal stirred gas phase process (Amoco, Chisso)

homopolymer reactor and high residence time in the EPR reactor, would be reduced [236–242].

In 1995, the cooperative agreement between Amoco and Chisso was ended.

6.7   Processes—Summary

The progression of PP manufacturing processes from primarily slurry to a broader mixture of processes is illustrated in Fig. 6.9, where the worldwide capacity is shown by process. Surprisingly, there is still a significant amount of PP manufactured with the slurry process, although most of those plants would use the more active supported catalysts to at least eliminate the catalyst residue removal. However, the vast majority of the new capacity has been in Spheripol and Unipol Processes, with another healthy share going to the other gas phase and bulk processes. This chart shows clearly how, although the Spheripol and Hypol processes led the expansion that began in the 1980s, the growth of the PP business has been enjoyed by all of the process licensors.

Some operating changes were needed with the use of more active catalysts in the fourth generation processes. First, because of the enormous quantities of monomer used by the catalysts, the level of poisons in the monomers needed to be reduced significantly. The chemistry of monomer poisons was studied, tolerable levels were determined, and measures for preventing or cleaning up occasional contamination of the monomer supply were taken.

The process and product also became more sensitive to specific changes in the catalyst composition. Therefore, the catalyst composition was more carefully controlled.

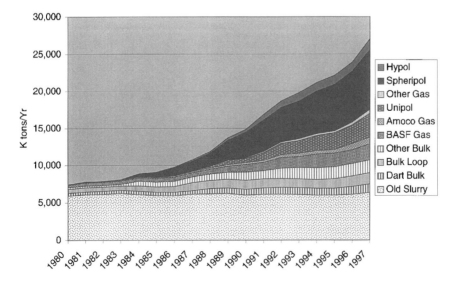

Figure 6.9   PP capacity growth by process

7   Product Developments

7.1   New Product Property Ranges

The expansion in PP product properties was not as dramatic as the explosion in catalysts or the evolution of new processes, but it added vigor to the business. Although the particular properties would depend on the catalyst used, the process, and the operating conditions, wider capabilities in properties were experienced, and are illustrated in Table 7.1.

Perhaps more important than the range of possible properties was the greater ability to adjust the operation to generate specific desired properties and special combinations of them.

The most beneficial property change was probably the increase in MFR that was possible; the former upper limit of about 35 g/10 min moved to about 1500. Thus, new PP grades allowed applications of thin-walled molded parts, thinner films, larger molded parts, and even finer fiber grades without melt degradation. Combinations of high melt flow in the homopolymer stage with tougher EPR allowed the creation of higher flow copolymers while retaining good low temperature impact behavior.

Many combinations of properties earlier considered impossible were soon entering the market. High flow/high impact copolymers, clear/good impact copolymers, very high flow/low oligomers were some examples of new regions that were accessible because of these new polymer capabilities.

7.2   Product—Process Conflicts

The attempts to expand the property range were not without difficulty. Most often, the acceptable limit to a new composition was determined by the process problems it would

Table 7.1   Property Range of 4th Generation Polypropylenes

Property	Units	Fourth gen. PP	Second gen. PP
Particle size	mm	0.3–5	< 0.5
Xylene solubles	%	90–99	92–95
Melt flow rate	g/10 min	0.2–1500	< 35
Melting point	°C	130–165	142–165
Flexural modulus	MPa	500–2400	1000–2200
Notched Izod impact	J/m	25–650	25–800

cause in manufacturing. Occasionally, manufacturing would be unaffected, but the customer encountered processing difficulties during fabrication. Because of the breadth of the changes arising from the fourth generation catalysts, difficulties were encountered in both areas.

Although the new catalysts expanded the range of operable conditions with respect to the earlier products, new products were constantly sought after, which, because of their unusual compositions, once again tested the limits of the manufacturing operation. A reliable operating point was often very difficult to define, especially in a large commercial operation. Fortunately, small but significant changes in the catalyst technology could sometimes expand the operating envelope enough to transform marginal conditions into a reliable operation.

As difficult as manufacturing problems were, the downstream difficulties, in the customers' operations, were a magnitude more intractable. This was even truer for the very large customers, such as the fiber and film producers, who operated complicated lines that had been fine-tuned to particular second generation polymers.

When the fourth generation polymers arrived on the scene, in the early 1980s, with known but small differences, the reception among the larger customers was less than enthusiastic. Perhaps the most crucial aspect was the MWD. In a typical fine denier fiber operation, the fiber extrusion occurs at a high temperature with a long residence time, and a considerable amount of melt degradation occurs. The result is a decrease in polymer intrinsic viscosity, and a consequent narrowing of MWD. The fiber spinning operation is very sensitive to both variables. Thus, when somewhat narrower MWD polymers, the normal result of fourth generation catalysts, were first offered to fiber producers, good operability was elusive.

However, at this point, the flexibility of the new catalysts also provided the solution. Once an effective dialogue was established with the customer, it was possible to tailor the new polymers, and the customer's operating conditions, to arrive at the same spinning performance enjoyed earlier.

Difficulties also occurred in BOPP film when the narrower MWD made the thickness control more difficult. The solution to that problem was more demanding in that, first, the relationship among polymer characteristics, film operating conditions, and final film thickness uniformity was not simple. Second, once having recognized the need for a broader MWD polymer, a modification of the polymer manufacturing process was required to achieve it.

Clearly, healthy supplier-customer relationships were essential for such achievements. While those were difficult times, the industry clearly overcame the problems.

7.3   New Applications

A few applications were the direct result of the products coming from the fourth generation catalysts. The familiar lawn chairs found in every sidewalk café and back yard patio worldwide simply could not have been molded at the quality and cost levels needed without the high melt flow rate products from this revolution. The growth of automobile bumpers in physical size, molded quality, and low temperature impact resistance demand PP behavior not formerly available. Thin-wall molded PP containers can compete with thermoformed alternative materials because of the current high melt flow rate capabilities. Bottle caps, with their intricate designs and high molding rates, are favored by the fourth generation developments. These are just a few highly visible examples of the revolution in PP at work. Section 9.2.3 describes the continuation of that progress beyond conventional PP products.

8 Commercial Developments

The broader range of PP properties available and the lower manufacturing costs placed PP in a more aggressive competitive position against other plastics, notably PE, PS, ABS, and to a lesser degree, PVC. While it is difficult to quantify that competition, the continuing high growth rate of PP was clearly stimulated by these developments.

The readily available technologies from several sources, with relatively simple plants, also opened the way for less technically oriented organizations to enter the PP business. The range of organizations that could confidently undertake a PP manufacturing venture was greatly broadened, just at a time when the less-developed countries of the world became hungry for local plastic production. Consequently, governments and companies in areas outside the "big three" (North America, Europe, Japan) began to install their own PP plants. One very important additional factor that allowed such expansion was the growth in propylene production in these areas, associated with new oil refineries [123]. Consequently, the rapid worldwide growth has been concentrated in new regions of the world. In Fig. 8.1, the growth of worldwide capacity illustrates the above trend. Clearly, the areas outside of the "big three" are not only catching up, but are rapidly exceeding the older producing regions. Considering that the consumption of

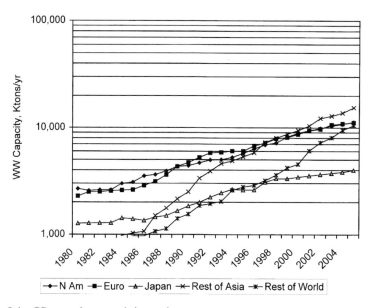

Figure 8.1   PP capacity growth by region

PP per capita is still about one magnitude less than in the "big three," this trend can be expected to continue quite strongly for the foreseeable future.

The simplification of the processes also lowered the barriers to manufacturing the more sophisticated products, formerly limited to the more technically capable producers. Consequently, the price differential for special products such as high impact copolymers (compared to homopolymers) has dropped. Thus, the competitive nature of the PP industry has intensified.

In summary, the industry enjoyed a revolution in catalyst chemistry, process economy, and product property range, to the delight of the world's PP customers. However, the producers, who initially enjoyed lower costs, are now finding the competition to be more intense than ever.

8.1   New Business Approaches and Alliances

8.1.1   Himont

By 1983, essentially all of the features of supported catalysts originally dreamed about within Montedison in 1970 were achieved or exceeded:

1. The phthalate/silane donor system allowed polymerization at an excellent level of isotacticity, while maintaining high activity
2. The chemical source support had been refined into a spherical support of narrow particle size distribution, produced in a continuous, reliable process
3. Conditions had been established in precontact and prepolymerization stages to maintain the polymer particle morphology, eliminating particle disintegration
4. The process adopted the optimum design approach, including elimination of catalyst residue removal and atactic polymer recovery, and, where appropriate, elimination of pelletizing extrusion, resulting in major savings in investment and operating costs
5. The flexibility of the process and catalysts allowed a wider range of product properties to be reached, improving the competitiveness of PP in existing markets, and creating opportunities in new markets.

Considering the limited progress that had been made by 1975, following the active $MgCl_2$ discovery in 1968, the significance and the pace of the discoveries and developments from 1975–1983 were no less than revolutionary. The catalyst, process, and product package was transformed from a technically interesting technology with promising potential, into an established technology, with enormous economic advantages and flexibility, including commercial demonstrations of reliable catalyst preparation and polymer production, plus a wider range of polymer properties.

The economics of the Spheripol process were very much as expected from Prof. Galli's 1970 projection: significant lower investment and operating costs, with essentially complete elimination of environmental problems from offstreams. The incentive to build plants internally, and to license the technology to other producers was extremely high.

While Montedison was completing the excellent work on the catalyst and process, another giant in PP, Hercules, was searching for a badly needed technology to upgrade its large but obsolete plant capacity. The polymer research people, who had recognized that Hercules probably did not have the capability to develop a competitive catalyst on their own, were instructed to keep abreast of new technologies that could be helpful in revitalizing the Hercules position. When some of the later developments at Montedison came to their attention, Hercules promptly began discussions about a possible joint venture with ME. Because Hercules held a large share of the world PP market, especially in the USA, the match between the ME technology and the Hercules markets was clearly attractive to both parties. The joint venture between the two, Himont, was formed in 1983, even as the final details of the technology were being demonstrated.

We know in retrospect how successful that undertaking was. After the formation of Himont in November 1983, following a review of the new process economics by the former Hercules executives, the order was given to install the new plants in North America as rapidly as possible. Any delay was considered "equivalent to throwing money away" [243].

In addition, it became possible to process the spherical form polymer particles in conventional downstream handling and extrusion equipment, with the savings associated with the elimination of the pelletizing extruders and their operation. In order to incorporate the additives that normally entered the polymer stream in the pelletizing extrusion operation, a process for topical application of stabilizers and other additives was developed by Himont, known as the Addipol process. The resulting processable spheres, designated Addiform, and marketed under the Himont (now Montell) trademark Valtec, became particularly attractive to processors desiring properties not available from earlier processes, such as very high melt flow rates.

However, some disadvantages hindered the use of the Addiform product. Homogeneity of additives within and between particles was not as high as in conventional pelletized products. Because of the lower bulk density, shipping costs were somewhat higher. In addition, the customer had to accept greater responsibility for achieving uniformity in the polymer melt, for adding special ingredients, and for addressing cleanliness and safety questions related to dust. Finally, the polymer, not having been though a melt filter, was not as clean as needed for critical applications such as fibers. As a result, the Addiform product was attractive primarily when there was a technical reason for its use.

At the time of forming Himont, Montedison's PE business was shifted to EniChem, although licensing of the catalyst technology for PE polymerization remained in Montedison. Later, the basic research facility at Novara became an EniChem property, and the remaining catalyst chemists originally with Montecatini were moved one more time, to Ferrara.

Not everything went smoothly in the implementation of the Spheripol process. A considerable amount of the North American market was in impact copolymers.

However, when the first copolymer plants were started in Lake Charles, LA, in 1985, fouling of the heat exchangers occurred, and reactor cooling was rapidly reduced to an inoperable level.

There were two schools of thought, with different approaches to the problem. One attributed the difficulty to the process, the plant layout, and the operating conditions; the other focused on possible difficulties with the catalyst.

The process group quickly conducted numerous trials. Their changes improved operations quickly, but not to an acceptable level. It was necessary to study the catalyst in some detail to arrive at the root of the problem, and to take appropriate corrective action. As a consequence of this search for the underlying causes of the operating difficulties, the copolymer plant operations were raised to an economically attractive point, and the Himont joint venture proceeded subsequently to generate outstanding earnings.

In early 1987, about 20% of Himont shares were offered to the public. In late 1987, partner Montedison purchased the shares held by Hercules. In 1990, ME acquired all publicly held shares, and Himont became an internal holding of Montedison.

8.1.2   Union Carbide: Unipol PP Process

As mentioned earlier, Union Carbide wanted to translate their success in PE with the gas phase Unipol Process over to the PP business. An agreement was reached with Shell Oil Company in Houston, TX, to construct a full-scale demonstration plant at Seadrift, TX, using the Shell catalysts and Carbide's process. Union Carbide operated the plant, and Shell marketed the polymer. The principal objective was to license the Unipol PP Process, which included impact copolymer capabilities.

The success of the venture is apparent from the growth of the Unipol PP Process capacity in Fig. 6.9.

As part of the FTC conditions for the formation of Montell in 1995, Shell divested their PP assets in the US, which were purchased by Union Carbide in 1996.

8.1.3   Sumika

In 1992, Phillips and Sumitomo announced an agreement to form a joint venture and manufacture PP in the US as Phillips Sumika Polypropylene Company, using a high yield catalyst in a gas phase process, both from Sumitomo. The plant, near Houston, TX, started up in 1996.

8.1.4   Targor

In 1997, Hoechst and BASF joined their PP assets in the joint venture Targor GmbH. This company thus became the largest PP producer in Europe, but is not active outside of Europe. Earlier, BASF had acquired the PP assets of ICI.

8.1.5   Montell Polyolefins

In April 1995, the PP and PE interests in Royal Dutch Shell, excepting those in the US, were joined with Himont to form Montell Polyolefins. Regulators in both Europe and the US carefully examined the effect of this venture on the competitiveness of the industry, and the new company was approved in both instances, with stipulations. Some Himont joint ventures and Shell's PP interests in the US were divested. As mentioned above, the latter were sold to Union Carbide, Shell's former partner in the US, in 1996.

The formation of Montell cured a fundamental Himont problem: limited monomer supply. Shell's surplus propylene (relative to its needs for PP) fit Himont's needs nicely.

8.1.6   Dow Chemical

Dow, long a major factor in several other plastics, and quite active in metallocene catalyst developments, has elected to join the ranks of conventional PP producers, using Montell's Spheripol process technology. Their plant in Europe went on-stream in mid-1998, and one in the USA is scheduled to start up in the year 2000.

Part IV

The Progress Continues

9   Continuing Technology Growth

9.1   Porosity—A New Dimension

9.1.1   Low Blush Copolymers

It was recognized in 1983, at the formation of Himont, that the Spheripol process, as then configured, with a single gas-phase reactor, was not capable of generating a "low-blush" copolymer. The need for "low-blush" copolymer in the USA was critical to Himont. This highly profitable product, widely sold by Hercules prior to the formation of Himont, polymerized HDPE of a rather low molecular weight within the EPR particles to reduce the tendency to stress-whiten (or, "blush") when stressed. It required a rubber content over 40%, and the production of HDPE in a reaction stage following the EPR reactor.

A third stage reactor (the second gas phase reactor) was added to the Ferrara pilot plant in December 1983 to pursue a process for this product. It was immediately apparent that insufficient polymerization occurred in the HDPE reactor, and effective operating conditions could not be found. The prospect of developing a more effective catalyst to overcome that inactivity was poor; years of effort would have been required. However, the need for low blush copolymer was immediate.

Again, the question was viewed differently from the catalyst and process points of view. Although the conventional wisdom regarded the loss of activity in the third stage reactor to be the result of normal chemical changes in the catalyst which caused decay in the reaction rate, the process people questioned whether the polymerization rate dropped for chemical or physical reasons. Within a short time, experiments revealed that the problem was due to physical, not chemical limitations. One possible way to relieve the physical limitation was to employ a higher porosity catalyst.

It was known from earlier spherical support work that dealcoholation increased the porosity of the support and the resulting polymer. The process group tried a highly dealcoholated support, and found that the resulting higher porosity allowed a strong increase in polymerization activity in the HDPE reactor. Once again, the Ferrara group had applied different approaches to a puzzling question, and came up with the needed answer. This time, it was the process approach that worked.

Although the new balance of polymerization rates in the homopolymer, EPR, and PE stages required some process adjustments, the preparation of low blush copolymers was optimized, and trial quantities were prepared in pilot plant operations in February 1985. The resulting polymer developed into a commercially successful reactor grade low-blush composition (See Section 9.2.1).

Thus, the means of adjusting the support and polymer porosity to the desired levels not only opened the way to low blush copolymers, but also, and perhaps more important, provided a new dimension to the Himont technological capabilities.

9.1.2   New Target: High Porosity

The realization that the porosity could be increased controllably with dealcoholation provided a new degree of freedom in polymer compositions. It was suggested in early 1984 that combining the more porous polymer from highly dealcoholated support with the long life catalyst from the fourth generation silane donors might provide a means of reaching the higher rubber content needed for even tougher impact copolymers.

9.1.3   High Rubber Copolymers

Meanwhile, the automotive industry was moving toward larger, tougher bumpers. Higher rubber content, higher melt flow rate heterophasic copolymers were desired. However, with conventional low porosity catalysts, both variables contributed to more difficult plant operations because of higher solubles. Again, in the 1984 and 1985 pilot plant trials, the more porous support was used successfully for a high rubber (50% rubber), high impact product suitable for the bumper application. In later developments, this result was translated into a higher melt flow rate version (See Section 9.2.1).

Thus, although later studies added to the understanding of porosity, the appreciation for its effect on the range of operable products was immediate, and it became a new segment of the technology package.

9.1.4   The Technology of Dealcoholation and Porosity

Work was conducted in 1984 on the effects of the dealcoholation process on the catalyst and polymer characteristics. With increasing removal of EtOH, the following occurred:

- Although the surface of the support was rougher, the morphology of the polymer improved, provided it was combined with adequate prepolymerization. This included smoother surface, and an increase in the fraction of spherical particles.
- Polymerization activity and polymer IsoIndex both decreased.
- The porosities of both the support and the polymer increased.

The morphologies of raw support (not dealcoholated) and a highly dealcoholated support appear in Figs. 9.1 and 9.2, respectively. The coarse porosity of the highly dealcoholated support is apparent.

The differences in polymer particle porosities are readily evident in Fig. 9.3: (a) low porosity, and (b) high porosity [244]. It is this porosity, visible in the light microscope, with pores greater than 1 micron in diameter, that accepts larger quantities of rubbery polymer or other special additives [102, 245]. Even with the highly porous catalysts, the

Figure 9.1   Morphology of raw support

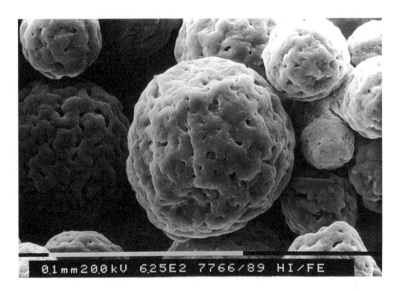

Figure 9.2   Morphology of highly dealcoholated support

replication of the catalyst shape was obtained in the polymer, provided the particles were protected from disintegration or attrition.

The 1984 studies also showed that the support became more brittle to ultrasound with increasing dealcoholation. However, instead of contributing to catalyst disintegration during polymerization, the polymers from the highly dealcoholated supports were

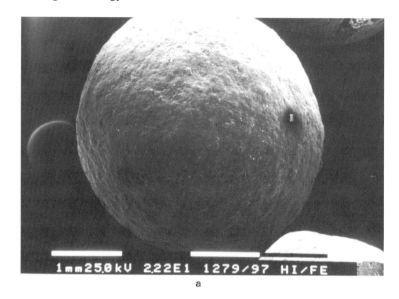

1mm25.0kV 2.22E1 1279/97 HI/FE

a

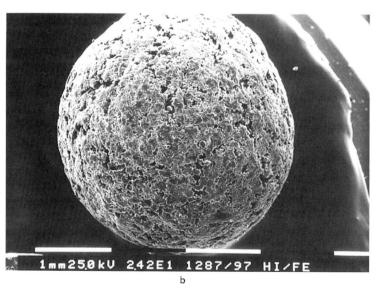

1mm25.0kV 2.42E1 1287/97 HI/FE

b

Figure 9.3   Morphologies of (a) low and (b) high porosity PP particles

tougher and showed better morphology, as already stated. Apparently the more brittle supports contributed to easier subdivision of the catalyst into the primary crystallites during polymerization. Additional studies, not described here, were needed to develop products with reduced friability, to survive normal processing without attrition.

The $MgCl_2 \cdot nEtOH$ complex goes through numerous poorly defined structures during dealcoholation. In some of these complexes, the oxygen of the EtOH occupies a position between the Mg and the Cl atoms [246, 247]. While not all the structures have been fully characterized, there is much evidence to indicate that at about 2 moles of EtOH per $MgCl_2$, the complex exhibits many aspects of a chain-like structure, similar to the brown β form of $TiCl_3$. Several other ethyl compounds (formate and acetate, for example) are also able to cause the development of the δ-form of $MgCl_2$ upon removal of the Et compound, and the complexes display similar structures to those of EtOH with $MgCl_2$ [248]. While these results provide fascinating insights into the behavior of the complexes, there still remains much to understand regarding the formation of the desired structure, the δ-form of $MgCl_2$, from the alcohol complex.

Following the discovery of the new dimension, porosity, the range of commercially available Montell catalysts was expanded to include the more porous supports, and the technology of porous catalysts and polymers was added to the Montell patent protection [245, 249, 250].

9.2   Reactor Granule Technology

The added dimension of porosity control provided a new degree of freedom in the versatility of the catalyst. Professor Galli, who had assumed the post of Vice President of Technology at Himont in 1985, moved to establish the potential that might be achieved by porous catalysts in a practical process. Beginning in 1987, extensive work on porous catalysts commenced. To clearly identify the new technology, he adopted the phrase, "Reactor Granule Technology" [251], defined as:

> Controlled, reproducible polymerization of olefinic monomers on an active $MgCl_2$ supported catalyst, to give a growing, spherical granule that provides a porous reaction bed within which other monomers can be introduced and polymerized to form a polyolefin alloy.

It soon became evident that, with this new technology, it was possible to expand the range of polymerization conditions and monomers far beyond those of conventional Ziegler–Natta catalysts, while maintaining processability associated with the outer polymer shell [252]. The most obvious example is the inclusion of a greater concentration of a difficult polymer such as a C_2–C_3 rubber, while maintaining processability.

Himont embarked on a new series of developments, many of which were not even envisioned just a few years earlier.

9.2.1   Commercial Products from Reactor Granule Technology

The first products made with this new technology involved modest changes in already existing products. The Reactor Granule Technology either made it possible to manufacture an otherwise inoperable product in the Spheripol process, or to modify polymer compositions, thus expanding the product properties range.

The low-blush copolymer type initially prepared in pilot plant in early 1985 (See Section 9.1.1) was scaled up to commercial operations. This product, originally of the old Hercules SB786 family, was quickly introduced into the North American plants following the use of a more porous, more dealcoholated catalyst. While the SB786 type of product was not new, the Reactor Granule Technology made it operable in the North American Spheripol process plants.

Second, the high rubber, high melt flow rate bumper material, also run in the early 1985 pilot plant campaign (See Section 9.1.3), evolved initially into the commercial product SP179, then into Hifax 7135. This latter material exhibits a melt flow rate of about 15, while retaining the excellent low-temperature impact behavior of the old composition, which had a MFR of 5. In addition to this major improvement in the impact/MFR balance, mold shrinkage and paintability had been enhanced by minor but significant changes in the rubber composition. This bumper material and the low-blush compositions represented the first modest expansions of the PP property envelope as a result of the higher support porosity. Great care was exercised in scaling up the new bumper composition, for it moved plant operations in a direction known to cause fouling and plugging. It was necessary to insure that the operating conditions employed provided an adequate safety margin, particularly in the full-scale plant operations. The eventual adoption of the SP179 type is testimony to the confidence in that result.

9.2.2   Gas Phase Process

The ability of a gas phase polymerization to tolerate higher levels of rubber without stickiness was established during the development of the Spheripol process (See Section 6.1). Even with the more porous catalysts, there remained a limit to the compositions that could be made in the Spheripol process without encountering stickiness. For example, low melting random copolymers were desired for the sealing layers in composite films. Because the random copolymers were prepared in the first, liquid monomer reactor stage, the solubles caused stickiness problems when the C_2 content exceeded about 4%, and limited the lowest seal initiation temperature (SIT) to about 120°C. Much of the solubles dissolved in the monomer and were deposited on the surface of the particle upon flashing of the monomer. Thus, for high ethylene random copolymers, a gas-phase polymerization was preferred. The solubles would remain inside the particle where they were created, and higher solubles (with higher C_2 content) could be tolerated before operating troubles were encountered.

Conversion of the pilot plant to two gas phase reactors in 1989 demonstrated the increased operating latitude. Random copolymers, actually a mixture of two different copolymer compositions, provided SIT's below 100°C, with acceptable operability and low solubles, as needed for FDA approval.

Several innovations were introduced into the gas phase process to provide better overall operation. The finishing section of the plant uses the same steaming and nitrogen drying arrangement as the Spheripol process, resulting in a polymer free of monomer, and giving a process without environmental problems from outflows.

Thus, the attempt to reach a wider range of properties, begun with the more porous reactor granule, advanced further with the use of a gas phase process, which become another piece of the technology package [253].

9.2.3   New, Unconventional PP Products

Once it was realized that major expansions in PP properties could be achieved by combining the Reactor Granule Technology with entirely gas phase polymerizations, numerous new product concepts were developed [254].

Following the development of soft copolymers, even more flexible copolymers were desired. The "Supersoft" compositions were developed, using random copolymers in the first stage in place of homopolymers. A rather wide range of Supersoft properties has been developed, based on the compositions of the two phases. Matches in MFR, modulus, surface tension, and crystallization behavior are important for the best balances in properties. These very soft copolymers were prepared with flexural modulus values around 70 to 120 MPa, while retaining satisfactory operability.

A bitumen replacement material, while a successful technical development, was not attractive commercially.

Adjustment of the refractive index or dimension of the dispersed phase to better match the continuous phase results in a copolymer with good optical properties and enhanced impact behavior, regarded as a "Clear Impact" material.

As mentioned above, the low seal initiation temperature (SIT) copolymers evolved through several phases. The difficulty was to obtain the low melting point without creating high levels of solubles, which presented problems with FDA approval. The most effective composition used a mixture of two different random copolymers.

Table 9.1   New Products from Reactor Granule Technology (1992)

	1st stage	2nd stage	
Type	Composition	Wt% of total product	Rubber composition
Soft copolymer	Homopolymer, or	10–40	PE
	low C_2 Raco	30–60	40–70% C_2 EPR or EPDM
Bitumen replacement	Homopolymer	50–90	15–45% C_2 EPR
Supersoft	Homopolymer, or	5–20	PE
	low C_2 Raco	40–80	< 40% C_2 EPR or EPDM
Clear impact	Raco	2–30	Special formula
Low seal initiation temperature	Raco*	35–70	2–10% C_2 to C_8
Superstif	Homopolymer, low MFR	40–90	Homopol, MFR > 50

*30–65% of total product, 2–10% C_4 to C_8 alpha-olefin.

For a Superstif material, the use of widely differing melt flow rates in the two stages provides sufficiently broad MWD, which, when combined with the nucleation that results from the high molecular weight polymer component, allows the attainment of substantial increases in modulus, but with a processable composition.

Table 9.1 lists the major new compositions that developed into commercial products [255–262].

9.3   The Catalloy Process

Once again, the concepts considered in the laboratory and pilot plant trials were transformed into a commercial plant design, aimed at providing a wide range of flexibility with respect to monomers, polymer compositions, and operating conditions. The process engineers soon drafted the "Catalloy" process: a multiphase, multi-monomer polyolefin process, employing three gas-phase stages to maximize the compositional flexibility, and to minimize the negative process effects of soluble materials. The process flow diagram is shown in Fig. 9.4.

Although only minor operations had been conducted at the micropilot plant scale, a bold corporate decision was made in late 1988 to build two commercial Catalloy process plants, one in Ferrara, and one in Bayport, TX. Such was the confidence of top management in the quality and reliability of the technical developments that the risk of proceeding directly to the commercial scale was considered to be minimal.

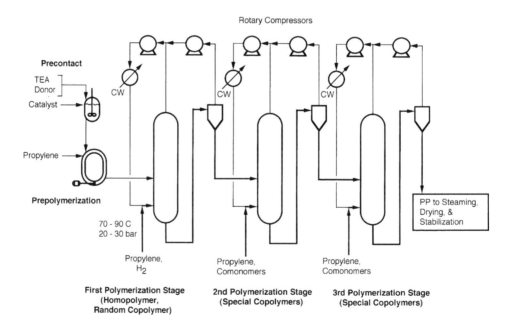

Figure 9.4   Catalloy process (Montell)

Table 9.2   Properties of Products from the Catalloy Process (1997)

Type	MFR	Flex Mod, MPa	Elong. at break, %	Hardness, D scale	Applications and other properties
Adstif KC 732P	20	2000	–	88[1]	Injection molding
Hifax 7135	15	950	>150	–	Bumpers; low shrinkage, good paintability
Hifax CA 53 A	10	650	>500	52	Injection molding
Hifax CA 138 A	3	420	>200	39	Masterbatch
Hifax CA 162 A	14	80	>500	32	Soft injection molding
Adflex Q 300 F	0.8	350	430	36	Extrusion, calendering
Adflex Q 100 F	0.6	80	800	30	Blown film, sheet
Adflex C 200 F	6	230	–	41	Coextrusion
Adflex X 101 H	8	80	–	–	Bitumen compounds
Adsyl 5 C 30 F	6	700	–	–	Seal init. temp: 105 °C

[1] R scale.

Some commercial products available in 1997 from the Catalloy process are illustrated in Table 9.2.

Because of the entirely new range of properties possessed by the products from the Catalloy process, many entirely new applications were found for these materials. The combination of new products, made by a new process, entering new markets (for polypropylene) increased the amount of time and effort required to establish commercial acceptance of these products. However, Montell has filled the original two Catalloy process plants, and a third was completed in 1997.

9.4   Further Donor Advances

9.4.1   New Silane Donors

Improved polymerization performance (higher activity and IsoIndex) remained a goal of the donor chemists. In addition, there was a desire to move away from aromatic groups, present in some donors, such as had caused the odor problems in 1978. In the early results, a new silane donor (Silane B), although still containing aromaticity, pushed the activity/IsoIndex limits to new levels, as seen in Table 9.3, and provided higher hydrogen activation at high MFRs, while retaining the low polymerization decay rate of Silane A.

Several new donors were created, Silanes C, D, and E, that, besides being aromatic-free, also gave higher activity and exceptionally high IsoIndex, in the range of desired melt flow rates, as indicated in Table 9.4 and Fig. 9.5. Curiously, however, especially for Silane E, the amount of hydrogen needed to generate a high melt flow rate was considerably higher. Therefore, these materials would not be very attractive for very high melt flow products.

Table 9.3   Silanes as 4th Generation Donors (Spherical Support/Liquid Monomer, 1987)

Catalyst	Activity, kg PP/g Ti	IsoIndex, %	Other attributes
Silane A	1870	96.3	Low decay rate; constant IsoIndex vs time and MFR
Silane B	2600	97.7	Higher activity at higher MFR

Table 9.4   Silanes as 4th Generation Donors (Spherical Support/Liquid Monomer, 1993)

Catalyst	Activity, kg PP/g Ti	IsoIndex, %	Other attributes
Silane A	1870	96.3	Low decay rate; constant IsoIndex vs time and MFR
Silane B	2600	97.7	Higher activity at higher MFR
Silane C	2500	98.0	Narrower MWD
Silane D	3400	98.8	High I.V., wide Al/Si range, high polym. temp.
Silane E	3500	99.0	Very broad MWD poor H_2 response

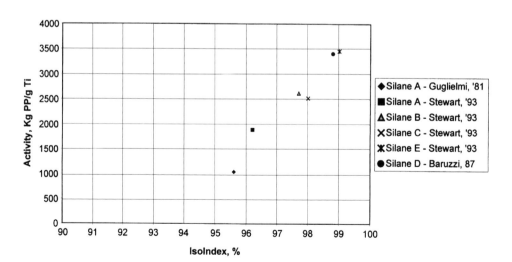

Figure 9.5   Performance of PP with silanes, in liquid monomer, spherical support

9.4.2   Diethers: No External Donor Needed

Encouraged by the success of the phthalate-silane combination, the catalyst chemists continued the search for more effective internal donors. That effort was driven by an extremely effective application of the principle that better understanding was the key to improved technologies. Molecular models of the active catalyst sites were used to determine the most effective steric arrangement for the donor candidates. Both the distance between donor atoms, usually oxygen, and the steric arrangement of the bulkier substituents were considered in this analysis. The phthalate esters were originally selected based on early work using this approach. Now the technique (Conformation Statistical Distribution) was applied rigorously to a large number of possible compounds, many never before synthesized [221].

While the phthalates were effective, another group, diethers, would be more resistant to reaction with the TEA, and could be synthesized to provide various distances between the two ether oxygens, and different appended chemical groups. Numerous structures were considered, and the more promising candidates were synthesized and tested. This resulted in the definition of a whole new class of donors, with appropriate patent protection [75, 114, 263–265]. Diketones were also found to have interesting activity, and were patented [266].

In addition to the activity/IsoIndex behavior shown in Table 9.5 and Fig. 9.6, the other outstanding aspects of this family of donors are [267]:

- Excellent performance even without the use of an external donor
- Higher Ti level (3.5–4.5%) with good IsoIndex
- Extremely high activity per gram of catalyst
- Very low cocatalyst consumption
- Very narrow MWD
- Very high MFR/H_2 response.

Table 9.5   Diethers as 4th Generation Donors (Spherical Support/Liquid Monomer, 1988)

Catalyst	Activity, kg PP/g Ti	IsoIndex, %	Other attributes
Silane A	1870	96.3	Low decay rate; constant IsoIndex vs time and MFR
Silane B	2600	97.7	Higher activity at higher MFR
Silane C	2500	98.0	Narrower MWD
Silane E	3500	99.0	Very broad MWD Poor H_2 response
Diether A[1]	1850	97.0	No external donor needed, narrow MWD, good H_2 response
Diether B[1]	4000[2]	96.3	No external donor needed, narrow MWD, good H_2 response

[1]Internal donor only.
[2]Activity per g catalyst is about double the best Silane results.

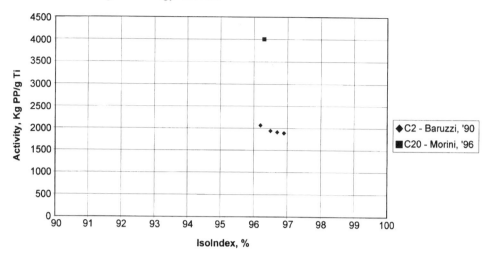

Figure 9.6   Performance of PP with diether donors

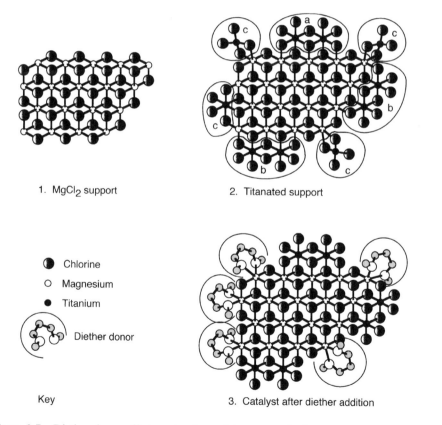

Figure 9.7   Diether donor effects—atomic model; see text for key

In addition to occupying the Mg sites not coordinated with a Ti atom, the diethers also remove single $TiCl_3$ groups, which are active atactic polymer sites. Next to corner sites, the diether converts the atactic site into a highly active isotactic site. Figure 9.7 depicts the three stages: (1) the $MgCl_2$ support, (2) the titanated support, and (3) after exposure to the diether [75]. In the titanated support, three types of $TiCl_4$ are shown: (a) Ti_2Cl_8 dimers that give isotactic polymer both before and after diether treatment, (b) Ti_2Cl_8 dimers that give atactic polymer before, then isotactic polymer, at a high rate, after diethers, and (c) $TiCl_4$ that gives atactic PP before, but are replaced by the diethers.

These donors are so different and improved over the phthalate/silanes that it has been suggested that they be classified as a whole new generation of catalysts [268], but we continue to classify them as fourth generation, although with exceptional properties. The diethers are expected to reach commercial operation in late 1998.

10 Expanding Technologies

The revolution is continuing beyond PP. The technologies that drove such immense changes in the PP industry are being applied to other promising areas. They are described below, only in enough detail to give some appreciation for the range of benefits coming from the supported catalyst developments. Perhaps the most important one is in metallocene catalysts, where the high promise of that technology has not yet been realized commercially.

10.1 Metallocenes

The metallocene catalysts are quite different than Ziegler–Natta catalysts. Each metallocene catalyst has characteristics specific to that structure. Consequently, it is possible to conduct polymerizations that have very specific properties. Distributions of molecular weight, copolymer composition, and stereospecificity are often extremely narrow. On the other hand, the range of stereospecificities that can be reached is quite wide: highly isotactic, atactic, and syndiotactic polymers are readily prepared, by using a different catalyst structure for each. There has been some variation in regioregularity, the 1,2 or 2,1 placement of the monomer in the growing chain, but recent catalysts have shown higher levels of regioregularity than earlier ones. Large quantities of cocatalyst, usually methylaluminoxane (MAO), are normally used in the metallocene polymerizations. One objective has been to eliminate the consumption of MAO to reduce the cost. Several companies have developed catalysts that are MAO-free.

10.2 Metallocene/Z–N Combined Catalysts

A greater difficulty is the inability to control the morphology of polymers from the metallocene catalysts. Here, PP from the Reactor Granule Technology can provide a solution. By dispersing the metallocene catalyst on a porous PP particle, the new polymer takes the shape of the PP. Thus, a wide variety of polymers, isotactic PP, syndiotactic PP, atactic PP, EPR, and the more sophisticated copolymers that are possible with the metallocenes can be produced with controlled morphology. The resulting slurry processability would be expected to approach that of the PP particles.

The metallocene catalysts are often regarded as the fifth catalyst generation for PP. The products obtained from the combination of the Ziegler–Natta catalysts and the metallocene catalysts offer so many new possibilities that it is expected to become the sixth catalyst generation.

10.3   Beyond Polyolefins: Hivalloy

Another innovative application of the Reactor Granule Technology is the introduction of non-olefinic polymers. Conventional catalysts for different types of polymerization may be dispersed on the porous PP particle, and either liquid or gaseous monomers are introduced and polymerized. A variety of non-olefinic polymers have been prepared at Montell in this way, allowing the new polymer, and often a graft copolymer, to be intimately dispersed within the PP particle. The sub-micron-sized dispersions are considerably finer than can be attained with extrusion mixing, and consequently, unusual combinations of properties are possible. These mixtures approach alloys in behavior.

Starting about 1989, an investigation using free radical-initiated styrene poly-merizations indicated that not only free polystyrene (PS) was forming, but also a PP–PS graft copolymer. This technology was expanded to maleic anhydride/styrene copolymers, acrylates, and compounded products of such grafted polymers with polyphenylene ether and other polymers [252].

These compositions, marketed as Hivalloy products, exhibit attractive combinations of high heat resistance, high impact, good flow, and good surface appearance [269, 270], and are thus carrying these PP-based products into markets well beyond those traditionally served by PP.

10.4   Return to PE: Spherilene

The catalyst technology that has produced so much versatility in the PP industry has also been re-examined for PE. A versatile PE process, called Spherilene, has been developed [271], and is being operated commercially.

11 Summary

With the return to its starting point, PE, it appears that the supported catalyst technology has "come full circle." The significance of the events during that 30-year trip has completely changed the world of PP.

The economic and scientific effects of these technological developments have been staggering. The polypropylene business has undergone a virtual revolution, both within itself, and with respect to competitiveness against other polymers. The distribution of PP production facilities around the world is rapidly spreading, particularly in the third world [123]. The range of materials we still call polypropylene has expanded substantially, increasing the markets being served by or being considered for PP. The earnings from PP sales, catalyst sales, and royalties from the licensing of these technologies have transformed Montecatini from a struggling Italian company in the 1960s into Montell, the largest company in the PP business worldwide, in the 1990s. Many of the old competitors and numerous new companies have adopted the supported catalyst in the PP business, with many variations on the basic technologies. Thus, the competition to manufacture PP is more intense, and the number of applications is continuing to expand.

These results were reached because a small cadre of technical leaders maintained their dream of continuing the flow of discoveries that began with Prof. Giulio Natta's 1954 discovery of crystalline PP. The results testify to the success of their quest.

Glossary of Terms

AlR₃	Trialkyl aluminum
Attrition	Break up of the catalyst or polymer particle into irregular shaped pieces due to processing, such as agitation or pumping
Bu	Butyl
CW	Cold water (used in flowsheets)
DIBP	Di*iso*butyl phthalate, a donor
Disintegration	Breakup of the growing polymer particle into irregular shaped pieces due to uneven polymer growth, usually occurring during rapid initial growth, with diffusion limitations
EB	Ethyl benzoate, a donor
EBR	Ethylene-butene rubber
ED	External donor
EPR	Ethylene-propylene rubber
EPDM	Ethylene-propylene-diene monomer terpolymer rubber
Et	Ethyl
EtOH	Ethyl alcohol
FT-1	A milled catalyst series developed by ME and MPC
GF	A precipitated catalyst. GF = granular form
Heco	Heterophasic copolymer
ID	Internal donor
II	IsoIndex
IsoIndex	Isotactic Index, also abbreviated II: the insoluble fraction after extraction with boiling heptane
LB	Lewis base
MRGT	Multi-catalyst reactor granule technology
MC	Montecatini
ME	Montedison
MPC	Mitsui Petrochemical
MPT	Methyl *para*toluate, a donor
PE	Polyethylene
PEA	*Para*ethyl anisate, a donor
PP	Polypropylene
PS	Polystyrene
PSD	Particle size distribution
PTBE	*p-tert*butyl ethyl benzoate, a donor
RGT	Reactor granule technology

Raco	Random copolymer
SIT	Seal initiation temperature
Subdivision	Division of the polymer particle into primary crystallites, embedded in the growing polymer particle, which ideally replicates the catalyst shape
TEA	Triethyl aluminum
TK	A precipitated catalyst developed by MPC
XI	Xylene insolubles
Z–N	Ziegler–Natta

Bibliography

1. *Boor, J. J.*: Ziegler–Natta Catalysts and Polymerizations (1979) Academic Press, New York.
2. *Quirk, R. P.*: Transition Metal Catalyzed Polymerizations: Alkenes and Dienes, Part A (1983) Harwood Academic Publishers, New York.
3. *Quirk, R. P.*: Transition Metal Catalyzed Polymerizations: Alkenes and Dienes, Part B (1983) Harwood Academic Publishers, New York.
4. *Kissin, Y. V.*: Isospecific Polymerization of Olefins (1985) Springer-Verlag, New York.
5. *Quirk, R. P.*: Transition Metal Catalyzed Polymerizations (1988) Cambridge University Press, Cambridge.
6. *Kaminsky, W. and Sinn, H.* (Eds): Transition metals and organometallics as catalysts for olefin polymerization (1988) Springer-Verlag, Berlin.
7. *Tait, P. J. T.*: In Comprehensive Polymer Science, G. C. Eastmond, et al. (Eds) (1989) Pergamon Press, p. 1.
8. *Keii, T. and Soga, K.* (Eds): Catalytic Olefin Polymerization (1990) Elsevier, New York.
9. *Van der Ven, S.*: Polypropylene and Other Polyolefins, Studies in Polymer Science (1990) Elsevier Science Publishers B.V., Amsterdam.
10. *Chung, T. C.*: New Advances in Polyolefins (1993) Plenum Press, New York.
11. *Corradini, P. and Guerra, G.*: Polym. Sci., Ser. A **36** (1994) pp. 1324–1340.
12. *Seymour, R. B. and Chang, T.*: History of Polyolefins (1986) D. Reidel, Dordrecht.
13. *McMillan, F. M.*: The Chain Straighteners (1979) MacMillan Press, London.
14. *Pino, P. and Moretti, G.*: Polymer **28** (1987) pp. 683–692.
15. *Natta, G. and Pasquon, I.*: Adv. Catal. **11** (1959) p. 1.
16. *Choi, K. Y. and Ray, W. H.*: J. Macromol. Sci. **C25 1** (1985) p. 57.
17. *Natta, G., Pino, P. and Mazzanti, G.*: Ital. Pat. 526101 (1956).
18. *Natta, G., et al.*: Gazz. Chim. Ital. **87** (1957) p. 549.
19. *Arlman, E. J. and Cossee, P.* J. Catal. **3** (1964) p. 99.
20. Germ. Pat. 2213086 (1972).
21. US Pat. 3769233 (1973).
22. US Pat. 4210738 (1980).
23. *Bernard, A. and Fiasse, B.*: In Catalytic Olefin Polymerization, T. Keii and Soga, K. (Eds) (1990) Kodansha Elsevier, New York, p. 405.
24. *Giannini, U., Zucchini, R. and Albizzati, E.*: J. Polym. Sci. **8** (1970) p. 405.
25. *Sittig, M.*: Polyolefin production processes—latest developments (1976) Noyes Data Corp., Park Ridge, NJ.

26. *Heggs, T. G.*: In, Ullman's Encyclopedia of Industrial Chemistry, B. Elvers, Hawkins, S. and Schulz G. (Eds) (1992) VCH Verlag GmbH, Weinheim, pp. 518–575.

27. *Lieberman, R. L. and Campbell, R. M.* (Montell): Personal communication: Mar 9, 1998, Wilmington, DE.

28. *Short, J. N.*: Rubber & Plast. Weekly (1980).

29. *Helm, C. D.*: US Pat. 3415799 (1968).

30. *Stryker, A. B., Jr. and Messina, P.* US Pat. 3462404 (1969).

31. *McCray, E. G.*: US Pat. 3554995 (1971).

32. *Stryker, A. B.*: US Pat. 3639374 (1972).

33. *Dietz, R. E.*: US Pat. 3318857 (1967).

34. *Buchanan, B. B.*: US Pat. 3342794 (1967).

35. *Hawkins, H. M. and Christiansen, D. C.*: US Pat. 3428619 (1969).

36. *Alleman, C. E.*: US Pat. 3324093 (1967).

37. *Trischmann, H.-G: et al.*: US Pat. 3652527 (1972).

38. Oil & Gas Journal (1970) Nov 23, p. 64.

39. *Trischmann, H. G., et al.*: US Pat. 4012573 (1977).

40. *Ross, J. F. and Bowles, U. A.*: Ind. Eng. Chem., Process Des. Dev. **24** (1985) p. 149.

41. *Hungenberg, K. D., et al.*: In, Ziegler Catalysts, G. Fink, Mülhaupt, R. and Brintzinger, H. H. (Eds) (1995) Springer-Verlag, Berlin, pp. 363–386.

42. *Hungenberg, K. D. and Kersting, M.*: In, New advances in polyolefins, T. C. Chung (Ed.) (1993) Plenum Press, New York, pp. 31–45.

43. *Hagemeyer, J. P. and Park, V. K.*: US Pat. 3423384 (1969).

44. *Hagemeyer, H. J., Hull, D. C. and Park, S. J.*: US Pat. 3600463 (1971).

45. *Hagemeyer, H. J. and Park, V. K.*: US Pat. 3679775 (1972).

46. *Simonazzi, T. and Giannini, U.*: Gazz. Chim. Ital. **124** (1994) pp. 533–541.

47. *Martuscelli, E., et al.*: Polymer (1985) 26 February, pp. 259–269.

48. *Del Duca, D. and Moore, E. P.* Jr., In, Polypropylene Handbook, Moore, E. P. Jr. (Ed.) (1996) Hanser Publishers, Munich, pp. 240–243.

49. *Boor, J. J.*: In Ziegler–Natta Catalysts and Polymerizations (1979) Academic Press, New York, pp. 158–166.

50. Belg. Pat. 650679 (1967).

51. GB Pat. 1024336 (1966).

52. *Delbouille, A. and Toussant, H.*: US Pat. 3594330 (1971).

53. *Delbouille, A. and Derroitte, J.*: US Pat. 3658722 (1972).

54. *Giannini, U., et al.*: Ital. Pat. 846003 (1969).

55. *Galli, P., DiDrusco, G. and Susa, E.*: US Pat. 3839313 (1974).

56. *Susa, E. and Mayr, A.*: Ital. Pat. 824213 (1969).

57. *Mayr, A., Susa, E. and Giachetti, E.*: US Pat. 4495338 (1985).

58. *Mayr, A., Susa, E. and Giachetti, E.*: US Pat. 4476289 (1984).

59. *Mayr, A., et al.*: US Pat. 4298718 (1981).

60. *Mayr, A.*: (Montell), Personal communication: April 12, 1996, Ferrara.

61. *Mayr, A., et al.*: US Pat. 4542198 (1985).
62. *Galli, P.*: Ferrara Technical report 53/70, Sept. 17, 1970.
63. Belg. Pat. 743325 (1969).
64. Belg. Pat. 744470 (1969).
65. Belg. Pat. 751315 (1969).
66. *Hock, C. W.*: J. Polym. Sci. **4** (1966) Part A-1, p. 3055.
67. *MacKie, P., et al.*: J. Polym. Sci., Part B **5** (1967) p. 493.
68. *Boor, J., Jr.*: Makromol. Chem., Rapid Commun. **2** (1967) p. 115.
69. *Galli, P., DiDrusco, G. and DeBartolo, S.*: US Pat. 3953414 (1976).
70. *Susa, E.*: Hydrocarbon Process. **51** (1972), p. 115.
71. *Kashiwa, N. and Tokuzimi, T.*: US Pat. 3642746 (1972).
72. *Boor, J. J.*: (1979) Academic Press, New York, pp. 112–115.
73. *Giannini, U.*: (Montell), Personal communication: May 28, 1996, Ferrara.
74. *Albizzati, E., et al.*: In Polypropylene Handbook, Moore, E. P. Jr. (Ed.) (1996) Hanser Publishers, Munich, pp. 22–45.
75. *Albizzati, E., et al.*: Makromol. Chem., Macromol. Symp. **89** (1995) pp. 73–89.
76. *Terano, M., Kataoka, T. and Keii, T.*: Makromol. Chem. **188** (1987) p. 1477.
77. *Mayr, A., et al.*: US Pat. 4544717 (1985).
78. *Mayr, A., et al.*: US Pat. 4636486 (1987).
79. *Albizzati, E. and Giannetti, E.*: US Pat. 4315836 (1982).
80. *Giannini, U., et al.*: US Pat. 4107413 (1978).
81. *Giannini, U., et al.*: US Pat. 4107414 (1978).
82. *Giannini, U., et al.*: US Pat. 4156063 (1980).
83. *Shiono, T., Uchino, H. and Soga, K.*: Polym. Bull. **21** (1989) p. 19.
84. *Nayak, P. R. and Ketteringham, J. M.*: In Break-throughs! (1986) Rawson Associates, New York.
85. *Kashiwa, N.*: US Pat. 3647772 (1972).
86. *Luciani, L., et al.*: US Pat. 4226741 (1980).
87. *Luciani, L., et al.*: US Pat. 4331561 (1982).
88. *Luciani, L., Barbé, P. C. and Simonazzi, T.*: Paper presented at ACS Centennial Meeting New York (1976).
89. *DiDrusco, G. and Luciani, L.*: Hydrocarbon Process. **60** (1981) p. 153.
90. *Matsuura, T.*: Chem. Econ. & Eng. Rev. **9** (1977) p. 28.
91. *Luciani, L.*: Paper presented at 37th Antec (1979).
92. *Luciani, L., et al.*: Paper presented at 40th ANTEC San Fransisco (1982).
93. *Bassi, I. W., Polato, F. and Calcaterra, M.*: Z. Kristallogr. **159** (1982) p. 297.
94. *Galli, P., et al.*: Eur. Polym. J. **19** (1983) p. 1977.
95. *Zannetti, R., et al.*: J. Polym. Sci. Part B: Polym. Phys. **26** (1988) p. 2399.
96. *Guidetti, G., et al.*: Eur. Polym. Journ. **16** (1980) pp. 1007–1015.
97. *Mazzullo, S.*: (Montell), Personal communication: Sept. 17, 1996, Ferrara.
98. *Barbé, P. C., Cecchin, G. and Noristi, L.*: In, Advances in Polymer Science 81 (1987) Springer-Verlag, Berlin.

99. *Galli, P., et al.*: Eur. Pol. J. **19** (1984) p. 19.

100. *Galli, P., Luciani, L. and Cecchin, G.*: Angew. Makromol. Chem. **94** (1981) No. 1441, pp. 63–89.

101. *Galli, P., Barbé, P. C. and Noristi, L.*: Angew. Makromol. Chem. **120** (1984) No. 1935, pp. 73–90.

102. *Simonazzi, T., Cecchin, G. and Mazzullo, S.* Prog. Polym. Sci. **16** (1991) pp. 303–329.

103. *Tait, J. T. and Abu Eid, M.*: Paper presented at 9th Symposium, Hibero-Americano der Catalyse Lisbon, Portugal (1984).

104. *Galli, P.*: Ferrara Technical report 197/76, June 1, 1976.

105. *Del Duca, D. and Moore, E. P. Jr.*: In, Polypropylene Handbook, Moore, E. P. Jr. (Ed.) (1996) Hanser Publishers, Munich, pp. 245–249.

106. *Cecchin, G. and Albizzati, E.* US Pat. 4294721 (1981).

107. *Cecchin, G. and Albizzati, E.*: US Pat. 4439540 (1984).

108. *Scatá, U., Luciani, L. and Barbé, P. C.*: US Pat. 4220554 (1980).

109. *Scatá, U., Luciani, L. and Barbé, P. C.*: US Pat. 4315835 (1982).

110. US Pat. 4401589 (1983).

111. US Pat. 4085276 (1978).

112. *Ferraris, M. and Rosati.*: US Pat. 4469648 (1984).

113. *Ferraris, M., et al.*: US Pat. 4399054 (1983).

114. *Albizzati, E., et al.*: In Ziegler Catalysts, G. Fink, Mülhaupt, R. and Brintzinger, H. H. (Eds) (1995) Springer-Verlag, Berlin, pp. 413–425.

115. *Albizzati, E.* (Montell), Personal communication: Sept. 25, 1996, Ferrara.

116. *Parodi, S., et al.*: Europe Pat. 45975 (1981).

117. *Parodi, S., et al.*: Europe Pat. 45976 (1982).

118. *Parodi, S., et al.*: Europe Pat. 45977 (1982).

119. *Parodi, S., et al.*: Europe Pat. 223010 (1981).

120. *Albizzati, E., et al.*: Makromol. Chem., Macromol. Symp. **48/49** (1991) pp. 223–238.

121. *Galli, P.*: (Montell), Personal communication: Feb. 11, 1997, Ferrara.

122. *Galli, P. and Spataro.*: US Pat. 4521566 (1985).

123. *Moore, E. P. and Larson, G. A.*: In Polypropylene Handbook, Moore, E. P. Jr. (Ed.) (1996) Hanser Publishers, Munich, pp. 257–284.

124. *Arzoumanidis, G. G., Khelghatian, H. M. and Lee, S. S.*: Euro Pat. 171179 (1986).

125. *Tovrog, H. S., Hoppin, C. R. and Johnson, B. V.*: US Pat. 4567155 (1986).

126. *Arzoumanidis, G. G. and Lee, S. S.*: US Pat. 4540679 (1985).

127. *Arzoumanidis, G. G., et al.*: US Pat. 4866022 (1989).

128. *Hoppin, C. R. and Tovrog, B. S.*: Euro Pat. 419249 (1991).

129. *Hoppin, C. R. and Tovrog, B. S.*: US Pat. 4829038 (1989).

130. *Karayannis, N. M. et al.*: In, Transition metals and organometallics as catalysts for olefin polymerization, W. Kaminsky and Sinn, H. (Eds) (1988) Springer-Verlag, New York, pp. 231–237.

131. *Chem. Week* (1992) Sep 10, 1992, p. 12.

132. *Staiger, G.*: US Pat. 4224184 (1980).

133. *Staiger, G.*: Euro Pat. 17895 (1980).

134. *Gruber, W. and Staiger, G.*: US Pat. 4579919 (1986).

135. *Gruber, W., Staiger, G. and Werner, R. A.*: Ger. Pat. 3411197 (1985).

136. *Zolk, R., Kerth, J. and Hemmerich, R.*: Euro Pat. 306867 (1989).

137. *Kerth, J., et al.*: Euro Pat. 288845 (1988).

138. *Kerth, J., Zolk, R. and Hemmerich, R.*: Euro Pat. 307813 (1989).

139. *Kersting, M., et al.*: Euro Pat. 655466 (1995).

140. *Kerth, J., et al.*: US Pat. 4977210 (1990).

141. *Schweier, G., et al.*: US Pat. 4745164 (1988).

142. *Warzelhan, V., Ball, W. and Bachl, R.*: US Pat. 4864005 (1989).

143. *Werner, R. A. and Zolk, R.*: US Pat. 4843132 (1989).

144. *Zolk, R., Kerth, J. and Hemmerich, R.*: US Pat. 4856613 (1989).

145. *Kelland, J. W.*: GB Pat. 371664 (1990).

146. *Cross, B. J., Bye, A. D. and Jones, P. J. V.*: Euro Pat. 290149 (1988).

147. *Cadlin, J. P., et al.*: Euro Pat. 37182 (1981).

148. *Gavens, P. D. and Caunt, A. D.*: Euro Pat. 14523 (1980).

149. *Kato, K., Nishimura, S. and Yokkaichi, M.*: Germ. Pat. 2951673 (1980).

150. *Kitigawa, S., O. I., and Saito, T.*: Germ. Pat. 2950437 (1980).

151. *Masakawa, S., Takahashi, T. and Yokoyama, M.* Germ. Pat. 2744559 (1978).

152. *Makukawa, S., Nimura, H. and Yoshida, S.* Ger. Pat. 3011326 (1980).

153. *Kono, M., Nimura, H. and Yoshida, S.*: Germ. Pat. 3230604 (1983).

154. *Yokoyama, M. and Sugano, T.*: US Pat. 4563436 (1986).

155. *Katou, K., Sugano, T. and Yokoyama, M.*: NL Pat. 86/01905 (1987).

156. Japan Pat. 87/246906 (1987).

157. Japan Pat. 88/277202 (1993).

158. Japan Pat. 89/001708 (1993).

159. *Fujii, M., et al.*: Euro Pat. 177841 (1986).

160. US Pat. 3642746 (1972).

161. GB Pat. 1554340 (1979).

162. *Kioka, M. and Kashiwa, N.*: Euro Pat. 22675 (1981).

163. Belg. Pat. 895019 (1983).

164. *Ushida, Y. and Kashiwa, N.*: Euro Pat. 86288 (1983).

165. *Kioka, M. and Kashiwa, N.*: Euro Pat. 125911 (1984).

166. *Ishimaru, N., Kioka, M. and Toyota, A.*: Euro Pat. 350170 (1990).

167. *Fodor, L. M.*: US Pat. 4260708 (1981).

168. *Selman, C. M.*: US Pat. 4246384 (1981).

169. *Welch, M. B.*: US Pat. 4330648 (1985).

170. *Welch, M. B. and Dietz, R. E.*: US Pat. 4331558 (1982).

171. *Goodall, B. L., van der Nat, A. A. and Sjardijn, W.*: Euro Pat. 19312 (1980).

172. *Goodall, B. L., van der Nat, A. A. and Sjardijn, W.*: US Pat. 4400302 (1983).

173. *Chadwick, J. C. and Ruisch, B. J.*: US Pat. 5061666 (1991).

174. *Chadwick, J. C. and Ruisch, B. J.*: US Pat. 5132379 (1992).

175. *Chadwick, J. C., Villena, A. and van Gaalen, R. P. C.*: Euro Pat. 357135 (1990).
176. *Villena, A., van Gaalen, R. P. C. and Chadwick, J. C.*: Euro Pat. 429128 (1991).
177. *Nestlerode, S. M., Burstain, I. G. and Job, R. C.*: US Pat. 4728705 (1988).
178. *Nestlerode, S. M., Burstain, I. G. and Job, R. C.* US Pat. 4771024 (1988).
179. *Job, R. C.*: US Pat. 4806696 (1989).
180. *Chadwick, J. C. and van der Sar, J. C.*: US Pat. 4663299 (1987).
181. Japan Pat. 84/142206 (1984).
182. Japan Pat. 85/161404 (1985).
183. *Terano, M., Soga, H. and Inoue, M.*: Euro Pat. 322798 (1989).
184. *Terano, M., Soga, H. and Inoue, M.*: Euro Pat. 459009 (1991).
185. *Terano, M., Soga, H. and Kimura, K.*: Euro Pat. 268685 (1988).
186. *Murai, A., et al.*: US Pat. 4847227 (1989).
187. *Flicker, H. K., Twu, H. C. and Liu, H. T.*: Euro Pat. 426140 (1991).
188. *Harada, M., et al.*: US Pat. 4551439 (1985).
189. *Miya, S., Tachibana, M. and Karasawa, Y.*: US Pat. 5100849 (1992).
190. *Uwai, T., Tachibana, M. and Hayashida, T.*: US Pat. 5122490 (1992).
191. *Harada, M., Iijima, M. and Saitoh, N.*: US Pat. 4499194 (1985).
192. *Band, E. I.*: Euro Pat. 113936 (1984).
193. *Takitani, M., Tomiyasu, S. and Baba, K.*: Euro Pat. 102503 (1984).
194. *Takitani, M., Tomiyasu, S. and Baba, K.*: Euro Pat. 103120 (1984).
195. *Takitani, M., Tomiyasu, S. and Baba, K.*: Euro Pat. 67416 (1982).
196. *Langer, A. W.*: US Pat. 4148756 (1979).
197. *Langer, A. W., Steger, J. J. and Burkhardt, J.*: US Pat. 4224182 (1980).
198. *Eur. Chem. News* (1981) Sep. 14.
199. *Diedrich, B. and Keil, K. D.*: US Pat. 3644318 (1972).
200. *Diedrich, B. and Dummer, W.*: US Pat. 3654249 (1972).
201. *Franke, R.*: Euro Pat. 302242 (1989).
202. Euro Pat. 68257 (1983).
203. Euro Pat. 249869 (1987).
204. WO Pat. 92/00332 (1992).
205. Euro Pat. 563815 (1993).
206. *Schreck, M. and Dolle, V.*: US Pat. 5100981 (1992).
207. *Iiskola, E. and Koskinen, J.*: WO Pat. 87/07620 (1987).
208. *Garoff, T., Leinonen, T. and Liskola, E.*: WO Pat. 9219659 (1992).
209. *Seppanen, H.*: Euro Pat. 262935 (1988).
210. *Koskinen, J. and Garoff, T.*: US Pat. 5468698 (1995).
211. *Baba, K., et al.*: Euro Pat. 72035 (1983).
212. US Pat. 4412049 (1983).
213. Euro Pat. 314169 (1989).
214. Euro Pat. 368344 (1990).
215. Euro Pat. 376084 (1990).
216. *DiDrusco, G. and Rinaldi, R.*: Hydrocarbon Process. **63** (1984) p. 113.

217. *Ferrero, M. A. and Chiovetta, M. G.*: Polym. Eng. Sci. (1991) 31 June, pp. 886–903.
218. *Hutchinson, R. A., Chen, C. M. and Ray, W. H.*: J. Appl. Polym. Sci. **44** (1992) pp. 1389–1414.
219. *Sacchetti, M.*: (Montell), Personal communication: Aug 8, 1997, Ferrara.
220. *Mei, G. and Cecchin, G.*: La Chimica e l'Industria (1996) 78 May, pp. 437–442.
221. *Barino, L. and Scordamaglia, R.*: Macromol. Symp. **89** (1995) pp. 101–111.
222. *Sacchi, M. C., Tritto, I. and Locatelli, P.*: Prog. Polym. Sci. **16** (1991) p. 351.
223. *Sacchi, M. C., et al.*: Macromolecules **25** (1992) p. 5014.
224. *Tait, P. J. T., et al.*: In Ziegler Catalysts, Fink, G., Mülhaupt, R., and Brintzinger, H. H. (Eds) (1995) Springer-Verlag, Berlin, pp. 343–362.
225. *Chadwick, J. C.*: In Ziegler Catalysts, Fink, G., Mülhaupt, R., and Brintzinger, H. H. (Eds) (1995) Springer-Verlag, Berlin, pp. 427–440.
226. *Choi, K.-Y. and Ray, W. H.*: Rev. Macromol. Chem. **C25** (1985) pp. 57–97.
227. *Simonazzi, T.* (Montell), Personal communication: Sept, 1996, Ferrara.
228. *Jorgensen, R. J., Goeke, G. L. and Karol, F. J.*: US Pat. 4349648 (1982).
229. *Hussein, F. D., Gaines, D. M. and Liu, H. T.*: Euro Pat. 251100 (1988).
230. *Brady, R. C., Stakem, F. G. and Liu, H. T.*: Euro Pat. 291958 (1988).
231. *Twu, F. C. and Burdett, I. D.*: Euro Pat. 341724 (1989).
232. GB Pat. 1044811 (1964).
233. *Jones, A. M.*: Paper presented at Polyolfins V International Conference Houston, TX (1987) p. 33.
234. *Shepard, J. W., et al.*: US Pat. 3957448 (1976).
235. US Pat. 4627735 (1986).
236. *Jezl, J. L., et al.*: US Pat. 3965083 (1976).
237. *Jezl, J. L. and Peters, E. F.*: US Pat. 3970611 (1976).
238. *Jezl, J. L. and Peters, E. F.*: US Pat. 4129701 (1978).
239. *Peters, E. F., et al.*: US Pat. 3971768 (1976).
240. *Stevens, J. F., et al.*: US Pat. 4326048 (1982).
241. *Brockmeier, N. F.*: Paper presented at Polyolefins VI International Conference Houston, TX (1991) p. 68.
242. US Pat. 4640963 (1987).
243. *Moler, E. S.*: (Himont), Personal communication: 1984, Wilmington, DE.
244. *Sgarzi, P.*: (Montell), Personal communication: Feb. 21, 1997.
245. *Govoni, G., Ciarrocchi, A. and Sacchetti, M.*: US Pat. 5231119 (1993).
246. *Valle, G., et al.*: Inorg. Chim. Acta **156** (1989) pp. 157–158.
247. *DiNoto, V., et al.*: Makromol. Chem. **193** (1992) pp. 1653–1663.
248. *DiNoto, V., et al.*: Makromol. Chem. **193** (1992) pp. 123–131.
249. *Sacchetti, M., Govoni, G. and Ciarrocchi, A.*: US Pat. 5221651 (1993).
250. *Govoni, G., Sacchetti, M. and Ciarrocchi, A.* US Pat. 5236962 (1993).
251. *Galli, P. and Haylock, J. C.*: Makromol. Chem., Macromol. Symp. **63** (1992) pp. 19–54.
252. *Galli, P., et al.*: Paper presented at IUPAC Akron, OH (1994).

253. *Covezzi, M., et al.*: Europe Pat. 541760 (1996).
254. *Cecchin, G.*: Macromol. Symp. **78** (1994) pp. 213–228.
255. *Cecchin, G. and Guglielmi, F.*: US Pat. 4302454 (1994).
256. *Cecchin, G., et al.*: US Pat. 5286564 (1994).
257. *Cecchin, G., Guglielmi, F. and Balzani, L.*: US Pat. 5077327 (1991).
258. *Cecchin, G. and Guglielmi, F.*: US Pat. 5298561 (1994).
259. *Covezzi, M.*: Euro. Pat. 483523 (1992).
260. *Pelliconi, A., Pellegati, G. and Vincenzi, P.*: Euro. Pat. 674991 (1995).
261. *Cecchin, G.*: Ital. Pat. MI 92A 001337 (1992).
262. *Cecchin, G., et al.*: Europe Pat. 573862 (1993).
263. *Albizzati, E. et al.*: US Pat. 4971937 (1990).
264. *Barbé, P. C., et al.*: US Pat. 4978648 (1990).
265. *Agnes, G., et al.*: US Pat. 5095153 (1992).
266. *Morini, G., et al.*: Europe Pat. 646605 (1995).
267. *Morini, P.*: (Montell), Personal communication: Nov. 19, 1996.
268. *Albizzati, E., et al.*: In Polypropylene Handbook, Moore, E. P. Jr. (Ed.) (1996) Hanser Publishers, Munich, pp. 11–14.
269. *Huang, M., Dong, D. and Wei-Berk, C.*: Paper presented at ANTEC '94 (1994).
270. *Dargis, K. R.*: Paper presented at SPO '96 Houston, TX (1996).
271. *Covezzi, M.*: Paper presented at Stepol Milan, Italy (1994).

Index

Ned Moore was born in Woburn, MA, on March 22, 1931. He attended schools in Winchester, MA, and obtained his B.S. in Chemical Engineering, Magna cum Laude, from Tufts University in 1952. He worked for Hercules and Himont, mostly in Wilmington, Delaware, from 1952 until 1993, except for two years in the Army from 1955 to 1957. Married in 1956, he and his wife, Georgene, have dedicated themselves to raising four children, from whom they derive both pleasure and pride.

After several years at the Hercules Research Center, Ned elected to remain in research. His involvement with PP began with the polymerization of polypropylene in a continuous loop reactor in 1959. In the following decade, he derived great satisfaction from the development of many new compositions and processes for biaxially oriented PP films, and became Manager of R&D for the Fiber and Film Department in 1969. He spent the late 70s managing the development of new resin compositions for polypropylene customers. In 1982, he became responsible for market development and management of the Specialty Products group. After Hercules joined Montedison to form the PP joint venture Himont, Ned returned to R&D in 1988 as a staff member of the Vice President of Technology. There, he became responsible for the worldwide flow and security of Himont's technical information, and helped manage the new technologies that issued from Himont.

Following retirement in 1993, Ned has consulted for Himont and Montell, primarily in a writing capacity. He was the editor of the "Polypropylene Handbook" published by Hanser in 1996, and has since provided consulting services to the Technology Company in Montell, including the writing of technical and historical documents. His latest publication, also through Hanser, is entitled "The Rebirth of Polypropylene: Supported Catalysts," which describes the revolutionary changes in the PP industry that resulted from the catalysts and processes developed within the Montedison laboratories in the 70s and 80s.

Ned and Georgie continue to reside in Wilmington, close to their two grandchildren.